Concepts in Applied Molecular Biology

CONCEPTS IN APPLIED MOLECULAR BIOLOGY

By

Ram Datar
Department of Pathology
Keck School of Medicine of the
University of Southern California
Los Angeles, CA, USA

with

Jenny Kim
Keck School of Medicine of the
University of Southern California
Los Angeles, CA, USA

BioTechniques® Press
Eaton Publishing
Part of the Informa Life Sciences Group

Ram Datar, MPhil, PhD
Assistant Professor of Clinical Pathology
Department of Pathology
Keck School of Medicine of the
University of Southern California
HMR 312C
2011 Zonal Avenue
Los Angeles, C A 90033, USA
datar@usc.edu

Jenny Kim, MS
Keck School of Medicine of the
University of Southern California
KAM 100-B
1975 Zonal Avenue
Los Angeles, CA 90033, USA
jjk@aludra.usc.edu

Library of Congress Cataloging-in-Publication Data

[CATALOGING IN PROGRESS]

ISBN 1-881299-30-9

Printed in the United States of America

9 8 7 6 5 4 3 2 1

Eaton Publishing
One Research Drive
P.O. Box 1070
Westborough, MA 01581-6070
www.BioTechniques.com

James D. Emerton: *Publisher*
Christine Emerton: *Managing Editor*
Alison Lozeau: *Scientific Copy Editor*
Sandy Lamont: *Production Manager*
Anne G. Hann: *Asst. Production Manager*
Erik Petersen: *Illustrator and Cover Design*
Robin Davis: *Marketing Projects Coordinator*

Contents

Major advances in the understanding of medicine and human disease are to a large degree dependent upon new technologies that allow biologists and medical scientists to envision and develop new concepts and cast aside the shackles of old and obsolete ideas. Approximately 150 years ago, a great leap forward occurred in medical sciences with the introduction of a practically useful microscope, coupled with the advent of the microtome for preparation of thin sections and a variety of biological stains. This microscopic revolution disproved the notion that diseases are caused by an imbalance of the "Four Humors" and introduced an entirely new concept based on disorders of tissues and cells. This in turn laid the groundwork for the advent of microbiology, for cellular descriptions of cancer, and the beginnings of immunology as a discipline. Such radical new technologies inevitably expanded the intellects of those who were willing to learn, and equally extended beyond the capability of the then-current vocabulary to describe new techniques and previously unimaginable findings. Thus, the lexicon of the medical scientist of 1900 would be largely unintelligible to his or her counterpart 100 years earlier, working prior to the advent of the era of cell biology.

A similar leap forward is occurring today with the fields of genetics, molecular biology, and genetic engineering extending our understanding and investigation of disease from the tissue and cellular level down to the level of individual proteins, genes, and even single moiety changes in DNA sequence, peptide sequence, or methylation status. Inevitably, these advancements have generated the need for a new lexicon to describe the variety of new methods and to document, in a clearly understandable manner, new findings in such a way that molecular biologists are able to communicate accurately across this rapidly changing field. This burgeoning vocabulary of course becomes familiar to each individual scientist, within his or her particular area of research, but the expansion of knowledge and technology has been so great that fields not immediately related to one's own area of expertise can quickly become difficult to comprehend, not so much because of the concepts involved, but rather because of the novel terminology that has evolved to describe the changes.

Concepts in Applied Molecular Biology is in one sense an illustrated medical dictionary, with a special focus on molecular biology, both as a field of discovery and as a series of interrelated research techniques. So fast has the field expanded that the ingenuity of scientists in developing appropriate terminology has been stretched almost to a breaking point, leading to such descriptors as "*trans*-splicing," "catalytic hammerhead ribozymes," "*Hin*dIII," "amplification refractory mutation systems," and "poly(A) tails," not to mention a bewildering array of acronyms, exemplified by SKY, SBH, DDRT-PCR, dNTP, and FISH. This new lingua scientifica is well understood by those who are immersed in the field, but to the rest of us is a cryptology for which we require a codebook.

Ram Datar and Jenny Kim have created that codebook. It is designed to serve as a rapid point of access and information for physicians and medical scientists within the field of medicine at large who are vitally interested in developing a sufficient understanding of this new field of molecular medicine such that we are able to integrate its findings and concepts into existing basic biological research, medical research, and medical practice. This book should therefore be invaluable to students of medicine and graduate students within medical schools, to practicing physicians, and to biological scientists and professionals in academic institutions and the biotechnology industry across the world.

Clive R. Taylor, MD, PhD
Professor and Chair, Department of Pathology
Senior Associate Dean for Educational Affairs
Keck School of Medicine
University of Southern California

Our motivations for writing this book were many: we felt that physicians, students of medicine and biology, biostatisticians, and even the now-essential research grant writers had a need to understand basic and advanced molecular biology concepts, and that the available texts in molecular biology provided either linear or branching tree-like information that is often too detailed and not necessarily helpful within a limited amount of time. Beyond that, the driving incentive behind this project was the desire to share our excitement over the discoveries made in molecular biology during the last two decades or so, and the picture of functional "wholeness" that is emerging.

The concepts described herein pertain to the research areas in which we have worked or studied and include some of the fields that are currently the most active. While the concepts and technical topics are arranged alphabetically for the sake of easier access to the reader, each topic can be read as a stand-alone entity, with suggestions for other related concepts. Thus, the reader can "start anywhere and go anyplace" in the book. The Linked Topics in the sidebars direct readers to related research areas, and the Suggested Reading lists at the end of each section offer readers the opportunity to delve into a given topic in more detail.

As Kary Banks Mullis, the inventor of PCR, once said: "Science consistently produces a new crop of miraculous truths and dazzling devices every year, truths and devices that enrich our lives . . . ." With an increasingly deeper understanding of the workings of molecular machines in living systems, what was truth yesterday may be replaced by a different truth tomorrow. By no means do we claim that any concepts described here are definitive truths; we are quite aware that some of the concepts as we understand them today will eventually evolve into substantially different ideas—in fact, we would welcome and embrace the expansion of knowledge in that direction in the near future, making it necessary to revise this book. We understand that obsolescence is built into the progress of molecular biology. But until that happens, we hope that this book will fill the lacuna that exists in the available literature, and allow bioresearch scientists and professionals who are curious about key concepts to find what they are looking for and know where to look for further information.

This book would not have been a reality without the constant inspiration (and sometimes, a push!) from a number of wonderful people. Ram would like to thank his persuasive wife, Bharati, herself a molecular biologist and often a valuable critic, and their son, Nakul, a budding computer engineer who taught Ram many word-processing tricks and ways to stay on course when things got hectic. He would also like to express his gratitude to his parents, Hemant and Mrinal, and parents-in-law, Gajanan and Sudha, for their patience and understanding throughout this process. Jenny would like to thank Ram for his advice, teaching, and friendship, and for his passion for science, which has been a constant inspiration for her. She would like to specifically and gratefully mention her brothers, James and Stanley, for their support, not just in writing this book, but also through life.

We would also like to express our appreciation for the efforts of Christine Emerton, Managing Editor of Eaton Publishing, who had remarkable faith in our ability and a single-minded focus on getting this work off our computer screens and into print. We are also grateful for the assistance provided by the production and marketing staff at Eaton, particularly Sandy Lamont and Robin Davis. Finally, this text would not be so informative without the detailed illustrations accompanying many of the topics; the extensive graphic work done by Erik Petersen has substantially improved the value of this book, and for this the authors are profoundly grateful.

Ram Datar and Jenny Kim
November 2003

Ram would like to dedicate this book to his grandfather, the late Dr. Madhusudan R. Datar, a compassionate physician and multilingual scholar, who believed so much in sharing.

Jenny would like to dedicate this book to her parents, David and Marie Kim, for their unending inspiration, faith, and support.

ALU REPEAT SEQUENCES

Linked Topics:

Cis-regulatory elements

PCR and PCR optimization

Physical mapping

Alu repeat elements are a family of short interspersed repetitive elements (SINEs) found exclusively in primates. These most common repeat elements are found in nearly one million copies per diploid genome. These repeats have a relatively high GC content. They are scattered throughout the human genome, yet have been shown to be preferentially located on the R bands in the Giemsa-stained chromosome preparations. The average frequency is 1 copy per 4 kb.

A typical full-length Alu repeat is about 280–300 bp long, flanked by an approximately 10-bp direct repeat sequence. It is a tandemly repeated dimer of an approximately 130-bp monomeric sequence, followed by a short sequence that is rich in A residues on one strand and T residues on the complementary strand. There is an asymmetry between the tandem repeats due to an internal 32-bp sequence present in one of the repeats that is absent in another.

Alu elements are dispersed by a mechanism called "retrotransposition." Three independent steps are involved in retrotransposition: transcription of the Alu repetitive element, reverse transcription of the Alu RNA, and integration of the Alu cDNA. There are two major Alu subfamilies, Alu-J and Alu-S. The Alu repeats are structurally related to the genes for the 7SL RNA component of the signal recognition particle, which facilitates the transport of proteins across the endoplasmic reticulum membrane. The 7SL DNA sequence family consists of 4 7SL genes, 500 7SL pseudogenes (which are truncated at one or both ends of the 7SL sequence), and 500,000 Alu sequences. Both 7SL genes and Alu elements are transcribed by RNA polymerase III. While transcripts from the far more abundant Alu elements are not readily detectable, RNA polymerase III transcribes the few 7SL genes efficiently. This is probably because most Alu elements and 7SL pseudogenes do not contain the first 37-nucleotide element upstream from the 7SL transcription start site, which is essential for efficient initiation.

Absent from coding sequences, Alu sequences are often associated with noncoding intragenic or intronic locations and, occasionally, in untranslated sequences. Because of their ubiquitous occurrence, Alu sequences are believed to promote unequal recombination, a mechanism that can be both pathogenic in some cases, but evolutionarily advantageous by promoting gene duplications. The Alu repeats may inactivate genes or alter their function and may, therefore, serve as a major driving force for evolution. Alu transcription can be activated by cellular stress, and Alu RNA might inhibit apoptosis through inhibition of the double-stranded RNA-activated kinase PKR. Alu demethylation in cancer might lead to a survival advantage through inhibited apoptosis.

Inter-Alu polymerase chain reaction (PCR), or Alu-PCR, is a variation of PCR in which Alu sequences are used as primers, and the regions of the genome between the repetitive Alu sequences are amplified. In the case of complex genomes like the human genome, such an interspersed repeat element-PCR (IRE-PCR) is used extensively for the subcloning of large clone inserts, such as those from yeast artificial chromosomes (YACs), and for screening the somatic cell hybrids and radiation hybrids for the presence of a human component.

Suggested Reading

Kolomietz, E., M.S. Meyn, A. Pandita, and J.A. Squire. The role of Alu repeat clusters as mediators of recurrent chromosomal aberrations in tumors. Genes Chromosomes Cancer *35*:97-112.

Batzer, M.A. and P.L. Deininger. 2002. Alu repeats and human genomic diversity. Nat. Rev. Genet. *3*:370-379.

Szmulewicz, M.N., G.E. Novick, and R.J. Herrera. 1998. Effects of Alu insertions on gene function. Electrophoresis *19*:1260-1264.

ANTISENSE OLIGONUCLEOTIDES

Linked Topics:

Aptamers

Gene therapy

Modified oligonucleotides

Oligonucleotide synthesis

Ribozymes

Most conventional drugs act by interfering with the function of enzymes or other proteins, but the precise mechanism of action is unclear in the case of the majority of drugs. The principal aim of pharmaceutics is the synthesis of rationally designed drugs with precisely targeted actions. Considering the fact that an actively transcribing gene can make up to 10,000 copies of mRNA, each of which can translate into 100 protein molecules, therapeutic intervention at the DNA or RNA level is certainly appealing as a more rational and effective way to impede the synthesis of a particular undesirable protein. Antisense oligonucleotides are a promising innovative class of such pharmaceutical agents, which can bind the target nucleic acids in a sequence-specific manner and bring about inhibition of disease-related genes.

Synthetic antisense oligonucleotides can bind the target nuclear DNA sequences by triplex formation via Hoogsteen base pairing, while they can also bind the target mRNA sequences via Watson-Crick base pairing. The former results in transcription inhibition, while the latter interaction can prevent mRNA splicing, polyadenylation, correct RNA folding, nucleocytoplasmic transport, translocation along the ribosomes, and translation initiation.

Zamecnik and Stephenson in 1978 first put forth the possibility of using antisense oligonucleotides to neutralize the viral nucleic acid targets. This approach has received a great deal of attention in the last decade. A wide variety of nucleic acid targets, such as viral genes, oncogenes, and genes for receptors, growth factors, ion channel proteins, enzymes, and others, have been tackled successfully in vitro. Using solid phase oligonucleotide synthesis, commercial DNA synthesizers can provide large quantities of antisense oligonucleotides. The effective dose of

Triplex DNA Formation

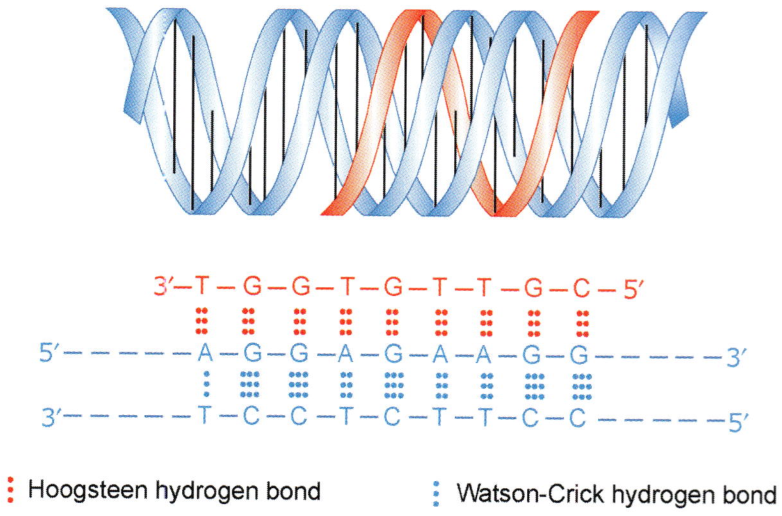

an antisense oligonucleotide for successful inhibition of transcription or translation is in the range of 0.1–50 μM.

The role of intracellular RNase H in antisense inhibition is controversial. This enzyme recognizes RNA–DNA hybrids and cleaves the RNA strand selectively, a role seemingly supportive of antisense translational inhibition. There is evidence, however, to show that DNA–RNA hybrids with only a few complementary base pairs can also serve as substrates for RNase H, thereby causing nonspecific cleavage of an mRNA; this could, in fact, be deleterious to the specificity of the antisense approach.

Although many of the synthetic antisense analogs are currently in phase III clinical trials, their actual in vivo therapeutic use will depend upon their eventual ability to satisfy the prerequisites of optimal cellular uptake, intracellular stability, target sequence specificity, and minimum toxicity. Moreover, in spite of the rational design, their ability to eliminate a single gene function in vivo has yet to be proven unequivocally. In order to achieve "true antisense" effects by increasing specificity, a variety of structural changes have been attempted to create modified oligonucleotides with better efficacy. In addition to ribozymes and peptide nucleic acids, the antisense approach signifies a plausible tool in future gene therapeutics.

Suggested Reading

Zamecnik, P.C. and M.L. Stephenson. 1978. Inhibition of Rous sarcoma virus replication and cell transformation by a specific oligodeoxynucleotide. Proc. Natl. Acad. Sci. USA *75*:280-284.

Dias, N. and C.A. Stein. 2002. Antisense oligonucleotides: basic concepts and mechanisms. Mol. Cancer Ther. *1*:347-355.

Kurreck, J. 2003. Antisense technologies. Improvement through novel chemical modifications. Eur. J. Biochem. *270*:1628-1644.

APOPTOSIS

Linked Topics:

Cell cycle

Genetic therapy

Multistep carcinogenesis

p53

The term apoptosis is generally used to describe programmed or cell-initiated cell death (Greek *ptosis* = shedding of leaves from trees). Apoptosis is distinguishable both morphologically and functionally from necrosis. Necrosis is a pathological cell death that occurs following physical or chemical injury, exposure to toxins, or ischemia (lack of oxygen). Characterized by swelling, rupture of the plasma membrane and cellular organelles, and release of the cellular contents into the surrounding tissue, necrosis leads to in vivo inflammation. Apoptosis, in contrast, is a natural form of death that organisms induce in response to physiological or developmental signals to remove cells.

The signal for apoptosis comes mainly from the "master switch": the p53 protein. In a dividing cell that has suffered DNA damage due to viral proteins, chemotherapeutic drugs or γ- and UV-irradiations, p53 recognizes the damaged integrity of DNA in the G_1 phase of the cell cycle, and (besides arresting the cell cycle in G_1 by activating the negative cell cycle regulator protein p21), forces the cell to undergo apoptosis by exerting a dual effect on the Bcl-2 family of apoptosis-related proteins. Thus, it represses transcription of the anti-apoptotic *Bcl2* gene, while simultaneously stimulating transcription of the pro-apoptotic *BAX* gene. Thus, p53 has been shown to be crucially involved in apoptosis. Cells lacking functional p53 fail to undergo apoptosis. Besides DNA damage sensed by p53, death signals also originate from contradictory, imbalanced, or insufficient growth signals from the microenvironment of the cell, as also from intracellular release of mitochondrial cytochrome C protein vital in cellular respiration.

Apoptosis is brought about by activation of an endogenous genetic program. While analysis of the process in the nematode *Caenorhabditis elegans* provided important clues, a number of genes participating in apoptosis are now identified even in mammals. Characterization of $t_{14/18}$ translocation found in the majority of the follicular B cell lymphoma cells led to the identification of the oncogene *Bcl2.* Overexpression of the Bcl-2 protein (which is now accepted to be a universal anti-apoptotic regulator) protects these cells from undergoing apoptosis. To date, at least nine members of the Bcl-2 family have been shown to function in the mammalian apoptotic pathway. Among these, Bcl-2 and $BclX_L$ (Bcl-like X protein, long form) are both anti-apoptotic in function due to their inhibitory action upon the pro-apoptotic proteins BAD (Bcl-2-associated death protein), BAK (Bcl-2-antagonist killer protein), BAX (Bcl-2-associated-X protein), $BclX_s$ (Bcl-like X protein, small form), and BIK (Bcl-2-interacting killer protein).

Upon receipt of the self-destruct signal, enzymes called procaspases are activated to caspases (also referred to as interleukin 1 β converting enzyme or ICE family members), which begin a cascade of activating a series of proteases, nucleases, and other catabolic enzymes. This initiation of caspase-mediated events is also brought about by cytochrome C

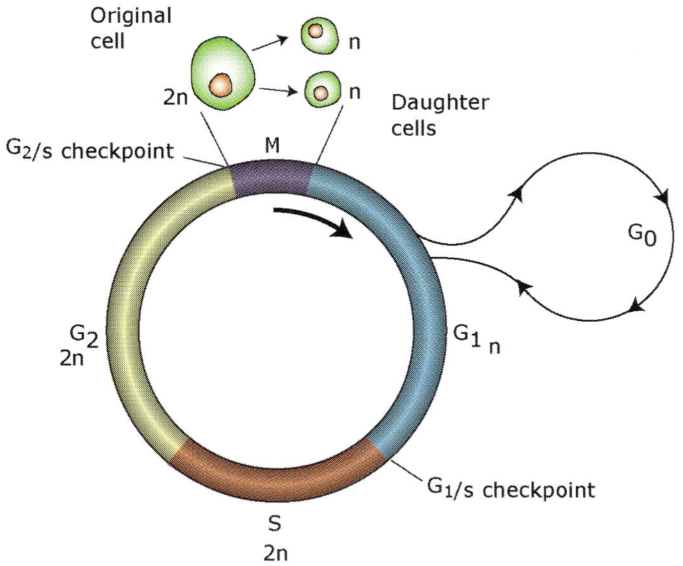

Stages of the Cell Cycle

M = mitosis, 1 diploid (2n) to 2 haploid (n) cells
S = DNA synthesis
G_0 = quiescent phase
G_1 = gap 1= Respond to extend stimulus to enter S phase or exit to G_0 in unfavorable conditions

G_2 = gap 2= Respond to stimulus allowing mitosis

released from mitochondria, which interacts with Apaf-1 (apoptotic protease activating factor-1) to activate procaspases. The nuclear chromatin condensation is followed by genomic DNA cleavage by the Mg^{++}- and/or Ca^{++}-dependent endonucleases resulting in formation of multimers of 180-bp-long fragments, observed on agarose gel electrophoresis as a characteristic DNA ladder. The cells dying by apoptosis usually shrink (but rarely lyse), and their membrane undergoes blebbing. Transglutaminase activity present at the onset of apoptosis, which cross-links proteins by creating peptide bonds through ε-δ glutamine-lysine bonds and transforms the normally fluid cytoplasm into a tight network, ensures that the integrity of the dying cell is maintained. The small vesicular membrane-bound products of cell fragmentation are recognized and efficiently engulfed by macrophages in the surrounding tissue, where the cellular debris fuses with the lysosomes and is completely digested, bypassing inflammation.

Apoptosis has come to occupy a central place in our understanding of the multistep carcinogenesis process. Various apoptosis inducers are being tested for cancer treatment both in traditional pharmaceutics and in genetic therapy.

Suggested Reading

Kiechle, F.L. and X. Zhang. 2002. Apoptosis: biochemical aspects and clinical implications. Clin. Chim. Acta *326*:27-45.

APTAMERS

Linked Topics:

Antisense oligonucleotides

Genetic therapy

Modified oligonucleotides

Oligonucleotide synthesis

Ribozymes

Experiments directed toward the understanding of interactions between oligonucleotides and proteins at the threshold of the 20th century yielded a surprising finding: oligonucleotides can bind unique protein domains tightly and can sequester and functionally neutralize the proteins. This is termed **protein epitope targeting**. More experiments were then designed that resulted in the development of a novel technology where the target protein-driven in vitro selection by exponential amplification (**SELEX**) could be achieved for RNA sequences.

Since the technology of SELEX is based on conformation-specific protein epitope-binding by the oligonucleotide, the individual oligonucleotides were named **aptamers** (Latin *aptus* = to fit). The specifically bound RNA or oligonucleotide sequence can be eluted from the proteins and sequenced to identify the recognition sequence. Transfer of large amounts of chemically synthesized aptamers into cells can result in specific binding to predetermined polypeptides, thereby blocking its function.

The random-sequence (degenerate) oligonucleotides used in SELEX assays for development of aptamers are obtained by oligonucleotide synthesis. Fully degenerate oligonucleotides can be synthesized by delivering 25% each of the four nucleotide monomers A, C, G, and T for each base position during synthesis. While this can generate an enormous number of sequence permutations (4^n, where n is the length of the oligonucleotide), the use of partially degenerate oligonucleotides is preferred, so that the concentration of individual oligonucleotides is not very low. Using the resulting mixture of oligonucleotides to screen for specific affinity to the target protein, a desired aptamer can be isolated and subsequently manufactured in large quantities.

It is possible for "natural" aptamers in vivo to modulate availability of specific proteins temporally and spatially within a cell, and utility of such synthetic agents has been documented in several systems. Thus, double-stranded phosphorothioate modified oligonucleotides are shown to regulate the availability of interleukin (IL)-2 protein in B and T lymphocytes. In another experiment, a 15-base DNA aptamer was shown to preclude thrombin from participating in blood coagulation in vitro, resulting in prevention of blood clot formation. Use of short nucleotide sequences as **decoys** to trap and divert transcription factors is also an application of aptamer technology.

Although still in its infancy, the field of aptamer development is likely to involve eventual genetic modification of the target cells in an organism and, therefore, ought to be considered a form of genetic therapy. Besides the transcription-inhibitory antisense oligonucleotides, antigene triplex molecules, and catalytic RNA (ribozymes), aptamer technology really forms a fourth component of antisense therapeutics.

Suggested Reading

Burgstaller, P., A. Girod, and M. Blind. 2002. Aptamers as tools for target prioritization and lead identification. Drug Disc. Today *7*:1221-1228.

AUTORADIOGRAPHY

Linked Topics:

In situ hybridization

Nonisotopic labeling and detection methods

Southern and Northern hybridization

Autoradiography is a procedure for localizing a radiolabeled compound within a solid sample by recording an image produced in a photographic emulsion. In molecular genetic applications, the solid sample often consists of size-fractionated DNA, RNA, or protein samples that are electrophoresed within a gel and either fixed and dried in situ, or transferred onto a surface of nylon or nitrocellulose membrane (as in Southern or Northern hybridization or Western blotting). Alternatively, they may be located within fixed chromatin or tissue samples mounted on glass slides, as in in situ hybridization.

Half-lives of the radioisotopes vary from a few days to a few years. Thus, ^{35}S has a half-life of 87 days, ^{32}P has a half-life of 14 days, and tritium has a half-life of 12 years. They also vary in their energy of emission. Thus, while ^{35}S has 0.167 meV energy of emission, the values for ^{32}P and tritium are 1.71 and 0.018 meV, respectively. Thus, among the three, ^{32}P is the highest energy radionuclide.

Photographic emulsion consists of suspension of silver halide crystals in a clear gelatinous phase. Following passage through the emulsion of β particles or γ-rays by the decaying radionuclei, the silver ions are converted to Ag atoms. The resulting latent image can be converted to a visible image once it is developed. Development is an amplification process, in which the entire silver halide crystal is reduced to give metallic silver. The fixing process subsequently results in removal of any unexposed silver halide crystals, giving an autoradiographic image that provides a 2-dimensional representation of spatial distribution of the radiolabel in the original sample.

Direct autoradiography involves placing the sample in intimate contact with an X-ray film, a plastic sheet with a coating of photographic emulsion. The radioactive emissions from the samples produce dark images on the developed film. Best suited for the detection of weak to medium strength β-emitting radionuclides, such as tritium and ^{35}S, this method is not suitable for high-energy β particles, such as ^{32}P, because such emissions traverse through the film, resulting in the wastage of the majority of the energy. Therefore, for such samples, intensifying screens, which are sheets of solid inorganic scintillator, are placed behind the film. Their use results in image intensification, because the emissions that pass through photographic emulsion are absorbed by the screen and converted to light, effectively superimposing photographic emission.

Indirect autoradiography is a modification in which the emitted energy is converted to photons of light by a chemical that is termed the scintillator or fluor. In fluorography, the sample is impregnated with the liquid scintillator. The energy transferred from the radioactive emissions causes the scintillator molecules to emit photons, thereby exposing the photographic emulsion. This method is mainly used to improve the detection of weak emitters, such as tritium or ^{35}S, inside a gel matrix,

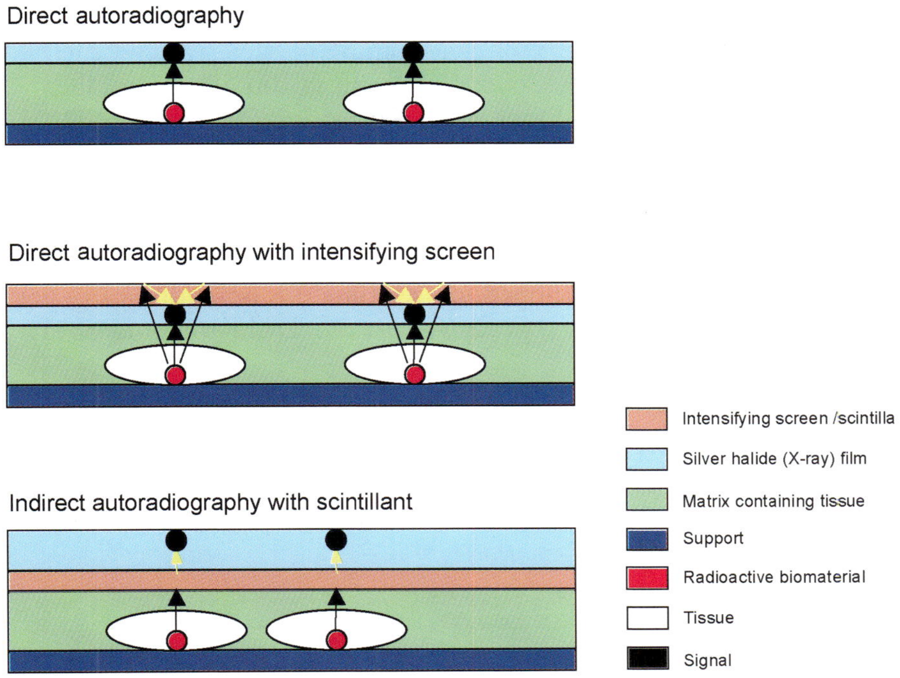

which otherwise absorbs a significant amount of the energy. The signal is enhanced when the gel is impregnated with a fluor, such as 2,5-diphenyloxazole.

Numerous radionuclides have been employed over the years to label biomolecules isotopically, and the technique of autoradiography has played an immensely important role in understanding fundamental processes in biology. Yet, increasing awareness of radiation hazards in recent years has led to the development of nonisotopic labeling methods and the gradual replacement of radionuclides with nonisotopic labels, such as digoxigenin, fluorescene, and biotin.

Suggested Reading

Link, A.J. 1999. Autoradiography of 2-D gels. Methods Mol. Biol. *112*:285-290.

Linked Topics:
Bacteriophages
Chromosome painting
DNA sequencing
Gene cloning methodologies
P1-derived artificial chromosomes
Physical mapping
Plasmids
Positional cloning
Pulsed field gel electrophoresis
Restriction endonucleases
Yeast artificial chromosomes

Many popularly used vectors for gene cloning in bacterial cells, such as plasmids and bacteriophages, contain high to medium copy number replicons. The advantage of such vectors is the increased yield of DNA they offer. However, the shortcoming is that such vectors often show structural instability of inserts, resulting in deletion or rearrangement of segments of cloned DNA. Such instability is particularly common in the case of large eukaryotic DNA inserts frequently containing repetitive sequences. Vectors developed to overcome this limitation carry low copy number replicons, such as *Escherichia coli* fertility plasmid or the F factor.

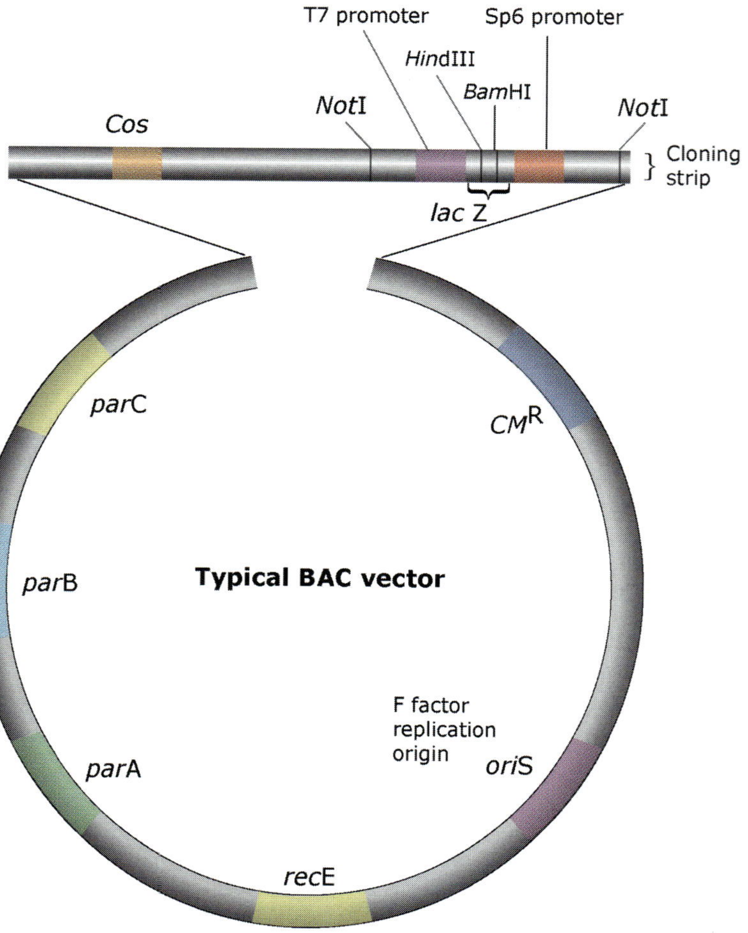

Developed as an alternative for yeast artificial chromosomes (YACs), the bacterial artificial chromosome (BAC) vectors carry six important elements: the F factor origin of replication (ori)S, a gene that dictates plasmid replication (*rec* E), and three genes that control maintenance of the copy number of the plasmid (*par* a, *par* b, and *par* c) at one to two per

E. coli cell. In addition, a typical BAC vector has a chloramphenicol resistance (*cm-r*) gene incorporated for antibiotic selection of transformants. The most popular BAC vector, pBeloBAC11, has three unique cloning sites for the restriction endonucleases *Hin*dIII, *Bam*HI, and *Sph*I within the *lacZ* gene. This permits identification of recombinants by colony color through α complementation. Combining the bacterial ori from BAC vectors with the yeast CEN6/ARS4 ori, the bacteria–yeast shuttle vectors are created. These vectors include a yeast *LEU2* gene for selection in yeast and the bacterial *cm-r* gene for selection in bacteria and are useful for cloning DNA stretches from YAC vectors.

The BAC vectors can accommodate as large as 300-kb-long foreign DNA fragments. The extraction of high molecular weight DNA for a BAC library is made with the insert DNA embedded in agarose. DNA is partially digested with restriction endonucleases, and fragments smaller than 150 kb are removed to reduce the number of BACs with small inserts. This is often achieved with pulsed field gel electrophoresis (PFGE). The vector is treated with a phosphatase to prevent self-ligation, followed by ligation to the selected genomic DNA fragments. The chimeras are transferred with high efficiency into bacterial cells using **electroporation**, a method of exposing cells to high voltage, which results in relaxation of the selective permeability of their plasma membrane. It has been shown, however, that the transformation efficiency is inversely related to the insert size.

Because the resulting BACs contain a low copy number replicon, only very low yields of recombinant DNA can be recovered from the host cells. This makes the isolation and screening difficult, necessitating large culture volumes to obtain sufficient quantities of BAC DNA for analysis, which, in turn, leads to a high RNA contamination. However, similar to P1-derived artificial chromosome (PAC) vectors, several columns and kits are commercially available for the isolation of BAC. Nevertheless, the evidence of rearrangement or recombination of the inserts in BAC vectors is practically nonexistent. BAC vectors also compare favorably with the YAC system due to the improved fidelity and greater ease of handling. These vectors, therefore, are important tools for positional cloning, long-range physical mapping, and whole genomic DNA sequencing. Another application of BAC vectors is their use in chromosome walking, where the strategy is to map the gene of interest down to a small segment of a chromosome and then stepwise arrange a series of overlapping clones contained in BACs or YACs.

Suggested Reading

Ball, K.D. and J.T. Trevors. 2002. Bacterial genomics: the use of DNA microarrays and bacterial artificial chromosomes. J. Microbiol. Methods *49*:275-284.

BACTERIOPHAGES

Linked Topics:
Bacterial artificial chromosomes
Gene cloning methodologies
In vitro packaging systems
Plasmids
Restriction endonucleases

Bacteriophages are viruses that infect bacterial cells. Bacteriophages with DNA as genomic nucleic acid often contain the DNA in double-stranded form, either circular or linear. Unlike plasmids that can exist extracellularly, the mature virus particles, or virions, have their genomes encased in a protein coat. This facilitates receptor-mediated adsorption onto a new host cell, where the DNA is then injected into the cell, leaving the protein capsid behind. The transmission of bacteriophage genomic DNA into specific bacterial hosts suggested the potential to use phage genomes as gene cloning vectors. The two most commonly used phage vectors are **bacteriophage lambda (λ)** and filamentous **bacteriophage M13**.

The wild-type bacteriophage λ has been the prototype for the analysis of a variety of different bacteriophages. It contains a genome of about 48-kb linear double-stranded DNA packaged within a protein coat. The termini of λ DNA have cohesive overhangs at the 5′ ends. Once inside the bacterial cells, these 12-nucleotide-long complementary ends pair with each other and form a cohesive site, *cos*, which plays an important role in late gene transcription and DNA packaging of λ.

The λ DNA can enter two alternative pathways inside a bacterial cell: the lytic cycle or the lysogenic state. In the **lytic cycle**, the λ DNA replicates initially bidirectionally, forming a θ-like structure, and subsequently by a rolling circle model, generating linear concatemers of the unit length. Core proteins such as terminase are synthesized that snip the λ concatemers at the *cos* sites to generate unit lengths of λ genome, which are packaged within the protein coats. The product of the λ gene *cII* brings about lysis of the host cell, allowing the progeny virions to escape and infect neighboring cells. In the **lysogenic state**, the λ genome takes recourse to its *att* gene. The *att* gene has a homolog in the *Escherichia coli* chromosome, and the proximity of the two *att* genes allows recombination between the λ and the bacterial genomes and subsequent integration of the λ DNA within the bacterial chromosome. λ DNA can remain in a stably integrated state (**provirus state**) for long periods, but may be excised from the host chromosome before entering into the lytic cycle. The decision between the lytic cycle or the lysogenic state is controlled by the λ regulatory genes *cI* and *cro*. These genes are functionally divergent. The dominance by the Cro protein leads to the repression of *cI* and onset of the lytic state, whereas the activation of the *cI* repressor suppresses the transcription of immediate early λ genes, including *cro*, and the lysogenic state sets in. In normally growing host cells (lysogens), the lysogenic state is favored, and the λ genome replicates along with the host chromosomal DNA.

Out of the 48-kb-long typical λ genome, the genes located in the central 18-kb segment (Stuffer segment) are dispensable and can be replaced by foreign DNA. Since the phage protein coat can accommodate DNA up to 110% of its genomic size, a foreign DNA insert up to

23 kb in length can be recombined in replacement cloning using appropriate restriction endonucleases. The insertion λ vectors, on the other hand, are generally used for cloning cDNA, which are smaller than 5 kb. Development of in vitro packaging systems has allowed attachment of the λ replicon to the target DNA and high efficiency of infection. Several major cloning systems have been derived from the original λ bacteriophage, including cosmids (plasmid vectors containing *cos* sequences), expression vectors (such as λ gt10 and λ gt11), and bacterial artificial chromosomes (BACs).

Suggested Reading

Campbell, A. 2003. The future of bacteriophage biology. Nat. Rev. Genet. *4*:471-477.

CELL HYBRIDS

Linked Topics:
Alu repeat sequences
Cell cycle
Human Genome Project
Modifications of fluorescence in situ hybridization
Modifications of PCR
PCR and PCR optimization
Physical mapping
Southern and Northern blotting and hybridization

Under the action of fusogens such as UV-inactivated Sendai virus or polyethylene glycol, cultured cells from different species can be induced to fuse together, thereby generating **somatic cell hybrids**. Thus, human cells can be fused with rodent cells, usually mouse or hamster cells, and the heterokaryons eventually proceed to mitosis, resulting in the dissolution of the two nuclear envelopes. The genomes of these hybrid cells are intrinsically unstable initially, and ultimately, most human chromosomes fail to replicate in subsequent cell cycles and are lost, giving rise to a variety of stable hybrid cell lines, each with a full set of rodent chromosomes plus one or a few human chromosomes. The loss of human chromosomes is random, and the human chromosome existing in somatic cell hybrids can be identified by karyotyping, taking advan-

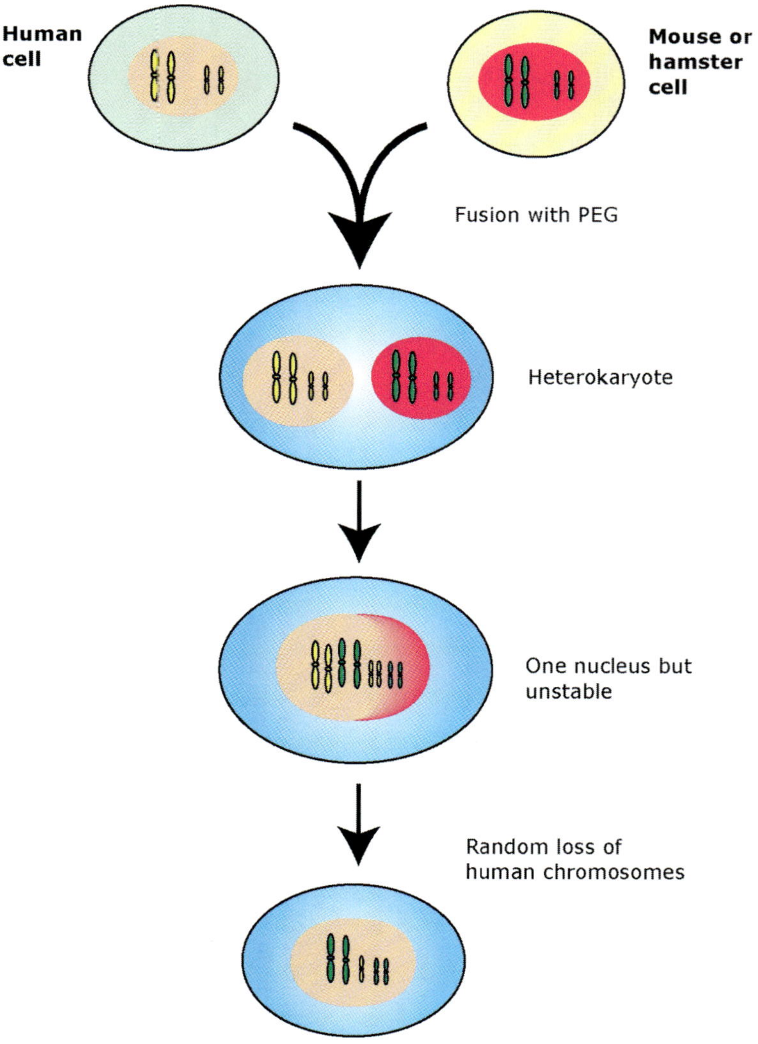

Stable Somatic Cell Hybrid

tage of the staining pattern of human chromosomes, which is distinct from that of rodent chromosomes. This can also be achieved by polymerase chain reaction (PCR) screening or by Southern blotting and hybridization with probes for Alu repeat restriction endonuclease cleavage fragments. The somatic cell hybrids can also be used to map human DNA sequences to specific chromosomal regions by fluorescence in situ hybridization (FISH).

Localization of a human DNA sequence to a single chromosome is inferred by deduction; that is, the target chromosomal region must be common to all those hybrid cell lines positive for the gene-specific probe and absent from the hybrid lines negative for the gene-specific probe. This is referred to as **concordant hybrids**. Problems due to the presence of multiple human chromosomes in traditional somatic cell hybrids are circumvented by technique of **microcell fusion**. This method takes advantage of the fact that exposure of the donor cells to a mitotic spindle inhibitor, such as colcemid, results in prolonged mitotic arrest and formation of numerous micronuclei. Centrifuging such mitotically arrested cells in the presence of cytochalasin B yields microcells, each harboring one chromosome. These can then be used to fuse with rodent cells, and the hybrids are called **monochromosomal hybrids**. These hybrids permit unequivocal evidence for the presence or absence of a targeted human DNA sequence at a specific subchromosomal region.

If human cells containing specific chromosome translocations or interstitial deletions are used as fusion partners with rodent cells, then somatic cell hybrids, called **translocation hybrids** or **deletion hybrids**, result. The deletions can also be created by treating the microcells with X-rays or γ-rays. Irradiated microcells with random deletions are fused with rodent cells to obtain **radiation hybrids**. Radiation hybrids can be used to order a set of DNA sequences or biochemically defined genes along a chromosome. Such ordering is feasible because, following irradiation and random DNA breaks, the probability of two target sequences being retained in the same hybrid is inversely proportional to the distance between them. Resolution of a radiation hybrid panel is determined by the extent of fragmentation and, hence, by the intensity of irradiation.

Taken together, the development of various somatic cell hybrids has aided the progress of efforts for gene mapping and the Human Genome Project considerably.

Suggested Reading

Kelsell, D.P. and N.K. Spurr. 1997. Gene mapping using somatic cell hybrids. Methods Mol. Biol. *68*:45-52.

Linked Topics:
Modifications of fluorescence in situ hybridization
Nonisotopic labeling and detection
Nucleic acid hybridization
Oligonucleotide synthesis
PCR and PCR optimization

Analysis of human metaphase chromosomes by classical karyotyping requires considerable expertise. Moreover, limitations of the technique (such as band-to-band similarities in different chromosomes, and interpreparation variation in uptake of stain by bands) make it difficult to quantify and automate the analysis of molecular cytogenetic anomalies. Such automation and quantification is made possible by employing individual chromosome-specific fluorochrome-labeled probes to "paint" the chromosomes. Chromosome painting is a special extension of fluorescence in situ hybridization (FISH) technology, which is particularly useful in defining the gene rearrangements in cancer cytogenetics, since chromosome preparations from tumors are often of poor quality. In association with the standard probes, chromosome paints can better define the complex chromosome arrangements so frequently seen in tumor samples. Chromosome painting can be applied to both metaphase and interphase nuclei.

Initially, the probes were prepared by combining all human DNA inserts from a flow-sorted DNA library of a specific chromosome. Later improvements saw the use of subchromosomal regions to yield to specific chromosome painting probes. These regions were obtained by microdissection subchromosomal DNA segments derived from chromosome-specific flow-sorted libraries or from somatic cell hybrids containing a single human chromosome. Currently, however, whole-chromosome probes for chromosomal painting are prepared by degenerative oligonucleotide primer-polymerase chain reaction (**DOP-PCR**) amplification of DNA. DOP-PCR is a PCR modification in which multiple oligonucleotides with partially random nucleotide combinations are used as primers. Thus, while preparing chromosome painting probes, about 120 combinations of the oligonucleotide sequence 5′-CCGACTCGAG-NNNNNATGT-GG-3′ (where N = any nucleotide) are used for primer-extension labeling at over 20,000 sites in each chromosome. This does away with the requirement of maintaining clones for chromosomes.

Figure 1. Fluorochromes

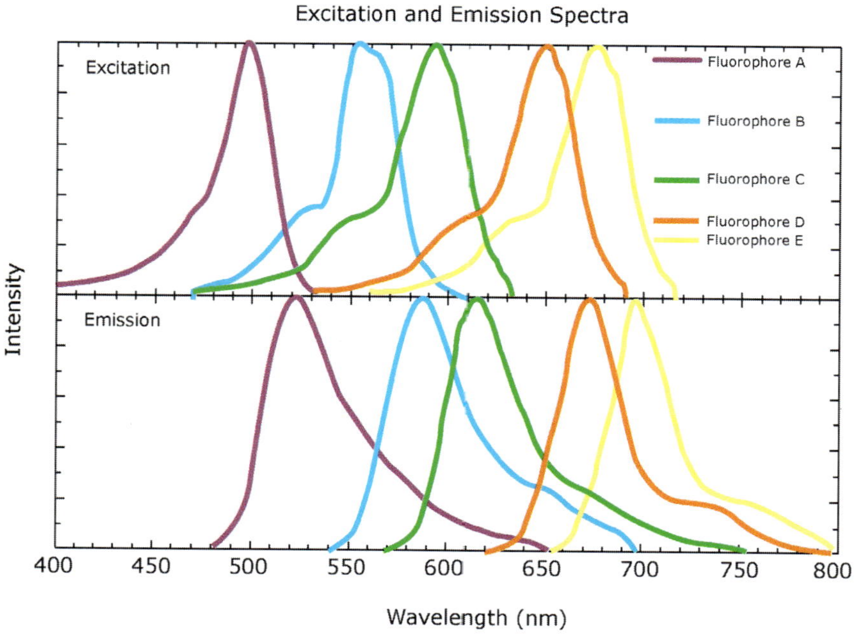

As of now, five synthetic fluorescent

labels are used in different ratios for nonisotopic labeling of the chromosomes. These fluorochromes have distinct excitation and emission spectra (see Figure 1). Variation of ratios allows expansion of the five basic spectral patterns to yield 24 differently fluorescent dyes required for visualizing each chromosome distinctly (see Table 1). The nucleic acid hybridization is performed essentially as in FISH.

Table 1. Fluorophore Permutation

Chr.	Fluorophore Permutation	Chr.	Fluorophore Permutation
1	BCD	13	AD
2	E	14	B
3	ACDE	15	ABC
4	CD	16	BD
5	ABDE	17	C
6	BCDE	18	ABD
7	BC	19	AC
8	D	20	A
9	ADE	21	DE
10	CE	22	ABCE
11	ACD	X	AE
12	BE	Y	CDE

A = Texas Red, B = Cy2, C = Cy3, D = Cy5, E = Ch5.5

Since the distinction between spectral properties of the labels can be undetectable to the human eye, specialized equipment and suitable image processing software are used to capture the fluorescent images and analyze the data, respectively. The program imparts pseudocolors onto the chromosome regions, presenting the data as a multicolor output. This technique, therefore, is often called multicolor FISH (**M-FISH**). Alternatively, taking into account the overall fluorescence emission spectrum from each entire chromosome, the program assigns a unique pseudocolor to each chromosome; this type of spectral karyotyping is called **SKY** (see Figure 2).

Figure 2. M-FISH

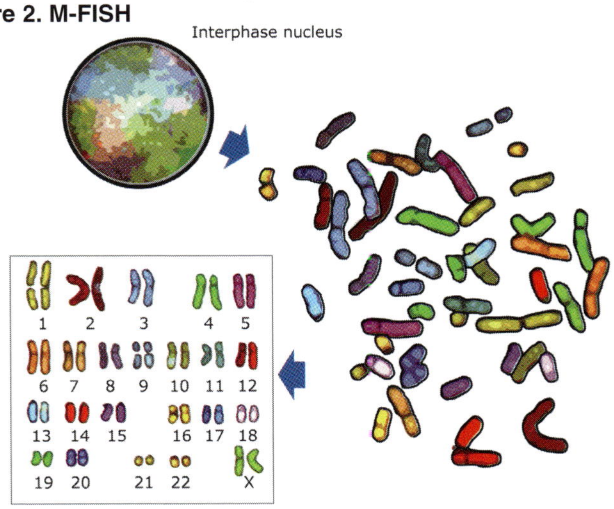

It is possible to use the chromosome painting, M-FISH, and SKY to analyze large chromosome deletions and amplifications and translocations between nonhomologous chromosomes. However, the current limitations include the inability to detect intrachromosomal inversions or duplications and small deletions. These limitations may be resolved as new dyes become available.

Suggested Reading

Sharma, A.K. and A. Sharma. 2001. Chromosome painting—principles, strategies and scope. Methods Cell Sci. *23*:1-5.

Fauth, C. and M.R. Speicher. 2001. Classifying by colors: FISH-based genome analysis. Cytogenet. Cell Genet. *93*:1-10.

Bayani, J. and J.A. Squire. 2001. Advances in the detection of chromosomal aberrations using spectral karyotyping. Clin. Genet. *59*:65-73.

Cis-REGULATORY ELEMENTS

Linked Topics:

DNA-binding proteins

Trans-*acting elements*

Eukaryotic genes have complex regulatory elements in the proximity of (or some distance apart from) the actual gene. The *trans*-acting elements, or transcription factors, recognize and bind specific *cis*-acting sequence elements (Latin *cis* = close or near). These *cis*-acting elements are about 8–10 nucleotides long and fall into four classes: promoters, response elements, enhancers, and silencers. Often, human genes and genes of other eukaryotes show evidence of three or more classes of *cis*-acting regulatory elements, each composed of multiple short sequence elements spread over a few hundred base pairs. Since these genomes are large, the chance occurrence of *cis* elements with the same sequence is higher.

Promoter: These are usually located in the immediate upstream regions of the gene, often within 200 bp from the transcription start site and serve to initiate the transcription by binding with RNA polymerase. They consist of combinations of short sequence elements, such as CCAAT (**CAT box**), **TATA box**, or GC box, which are recognized by specific transcription factors. The promoters for genes that show tissue-specific expression patterns additionally possess *cis*-acting elements that can be recognized by transcription factors specific for those tissues.

Enhancers: These are positive regulatory elements which, in contrast to the promoters, are independent of their orientation and their distance from the genes that they regulate. They contain elements recognized by both ubiquitous and tissue-specific transcription factors. The transcrip-

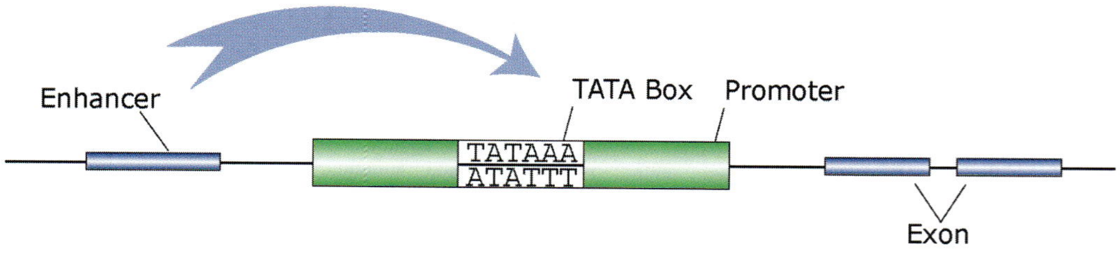

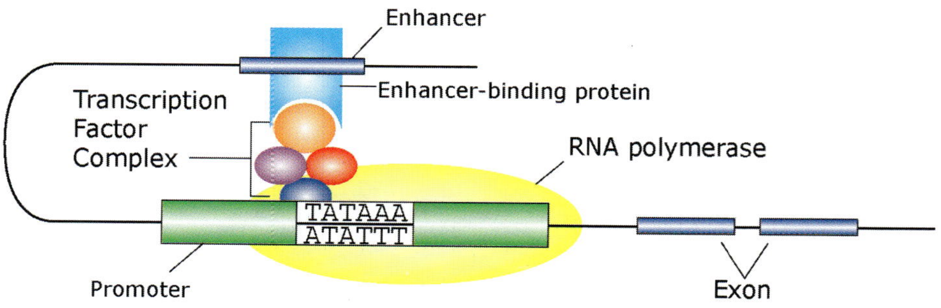

tion factors bind to such enhancer sequences, and the intervening DNA sequence between the enhancer and the promoter loops out. This brings the enhancer sequence bound to transcription factors in the vicinity of promoter sequences bound with RNA polymerase, which enhances transcription initiated by the promoter elements.

Silencers: In vitro DNA binding studies have shown the existence of these negative regulatory elements close to the promoter, but their significance in vivo is not clear.

Response elements: These are found only in selected genes, the expression of which is controlled by external factors such as hormones or growth factors, or by internal signaling molecules such as cAMP. These sequences are located a short distance upstream of the promoter elements, usually within a thousand nucleotides of the transcription initiation site. A wide variety of response elements for factors like glucocorticoids, retinoic acid, growth factors, cAMP, and others, are documented in different genes. The inducible genes are transcribed at high levels following the binding of the signaling factor to the corresponding response element.

Besides the above-mentioned elements, positive *cis*-acting regulatory elements may occasionally coordinate the expression of genes in a **gene cluster**, examples of which are found located 50–60 kb upstream of the genes for α and β globin gene clusters for binding with specific transcription factors, such as GATA I.

Suggested Reading

Urnov, F.D. 2003. Chromatin as a tool for the study of genome function in cancer. Ann. NY Acad. Sci. *983*:5-21.

Felsenfeld, G., J. Boyes, J. Chung, D. Clark, and V. Studitsky. 1996. Chromatin structure and gene expression. Proc. Natl. Acad. Sci. USA *93*:9384-9388.

COMPARATIVE GENOMIC HYBRIDIZATION

Linked Topics:

In situ hybridization

Modifications of fluorescence in situ hybridization

Multistep carcinogenesis

Alterations in oncogenes and suppressor genes significantly contribute to the multistep carcinogenesis process. A number of cancers are characterized by amplification in number of copies of structurally normal oncogenes. Amplifications in *erb2* and c-*myc* genes in breast cancers, the N-*myc* gene in advanced neuroblastomas, and the c-*myc* gene in Burkitt's lymphoma are a few of such examples. These extra copies can be intrachromosomal and show up as **homogeneously stained regions** (HSRs) when stained with Giemsa. Alternatively, they can be seen as small independent extrachromosomal entities, called **double minutes.** In contrast, cancer cells are also known to possess major deletions or losses in the suppressor genes.

The gene amplifications are often seen even in noncancer cells as a result of exposure to strong selective drug regimens. This phenomenon is exemplified by the amplification of the dihydrofolate reductase gene in cells that have developed resistance to methotrexate.

The technique of comparative genomic hybridization (CGH) uses a mixture of DNA from matched normal and tumor cells in competitive fluorescence in situ hybridization (FISH). The tumor DNA and normal control DNA are labeled with different fluorochromes and then hybridized in situ in equal quantities to chromosomes of normal cells in the presence of C_0t-1 DNA. C_0t-1 DNA is the sheared eukaryotic genomic DNA rich in repetitive sequences used as a competitor to block—and remove from hybridization reaction—the repetitive DNA sequences such as Alu repeat sequences in the target, minimizing the nonspecific hybridization of the probe to the repeat sequences. Besides CGH, C_0t-1 DNA is also used in microarray technology, Southern hybridization, and in situ hybridization.

When both the normal DNA and tumor DNA bind the template normal DNA in equal amounts, a resultant fluorochrome color pattern is seen. An excess fluorescence of tumor DNA type indicates amplification of that region in the tumor DNA, while predominance of a fluorescence from the normal DNA label indicates loss of the corresponding DNA sequence from the tumor DNA. The gain is indicative of oncogene amplification, whereas the loss suggests deletions, or aneuploidy, of tumor suppression genes. In principle, CGH can reveal all the regions of amplification and aneuploidy in a single experiment. Some of these alterations can be hitherto unknown deletions or amplifications, which can be studied further to identify new disease-related genes.

Suitable image processing software can be used to identify chromosomal regions where the ratio of fluorescent signal from normal and tumor DNA is different from the expected value. The regions of amplification or of losses in the tumor DNA can be detected depending on the directionality of deviation. The technique of CGH has detected a variety of such gene amplifications and losses.

CGH Methodology

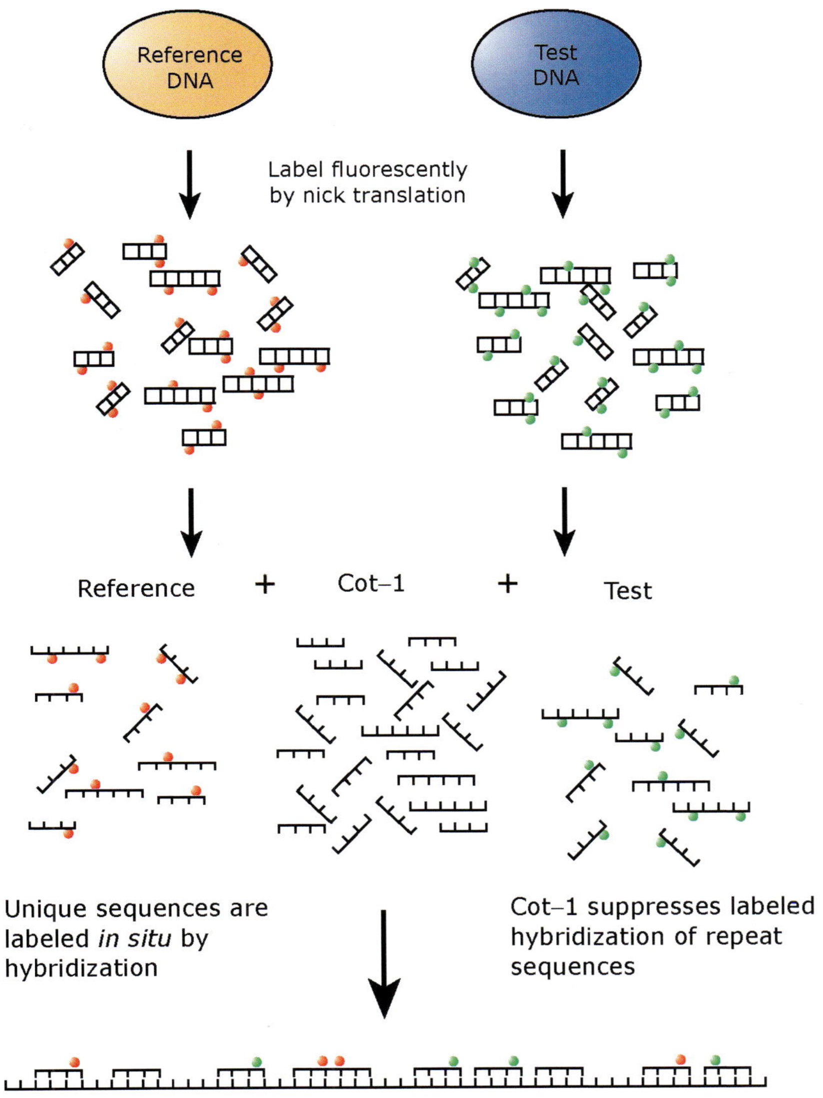

Suggested Reading

Lichter, P., S. Joos, M. Bentz, and S. Lampel. 2000. Comparative genomic hybridization: uses and limitations. Semin. Hematol. *37*:348-357.

Rowland, J.M. 2002. Molecular genetic diagnosis of pediatric cancer: current and emerging methods. Pediatr. Clin. North Am. *49*:1415-1435.

Oga, A., S. Kawauchi, H. Izumi, L.X. Ping, T. Furuya, and K. Sasaki. 2002. New perspectives for tumor pathology provided by comparative genomic hybridization. Int. J. Clin. Oncol. *7*:133-137.

Linked Topics:
Gene therapy
Molecular chaperones
Trans-*acting elements*
Two-hybrid systems

Transcriptional activators and transcriptional repressors are two types of regulatory proteins that act in concert to fine-tune the delicate balance of expression and silencing of genes. Both these proteins have DNA-binding domains through which they recognize and bind particular DNA sequences in the vicinity of the coding region of the target gene. Although the mechanism of action of eukaryotic repressors is not fully understood, a tremendous amount of work has been done on the nature and activity of transcriptional activators in prokaryotes and eukaryotes, and certain ground rules have been defined.

An activator protein binds to the specific DNA sequence near the target coding sequence. By virtue of additional domains that dock with proteins forming the transcription machinery (which include RNA polymerase), the activators turn on the transcription of the gene. Compared to 5-subunit RNA polymerase complexes in bacteria, the eukaryotes have about 50 proteins besides RNA polymerase forming the transcriptional complexes that are assembled prior to gene induc-

DNase1

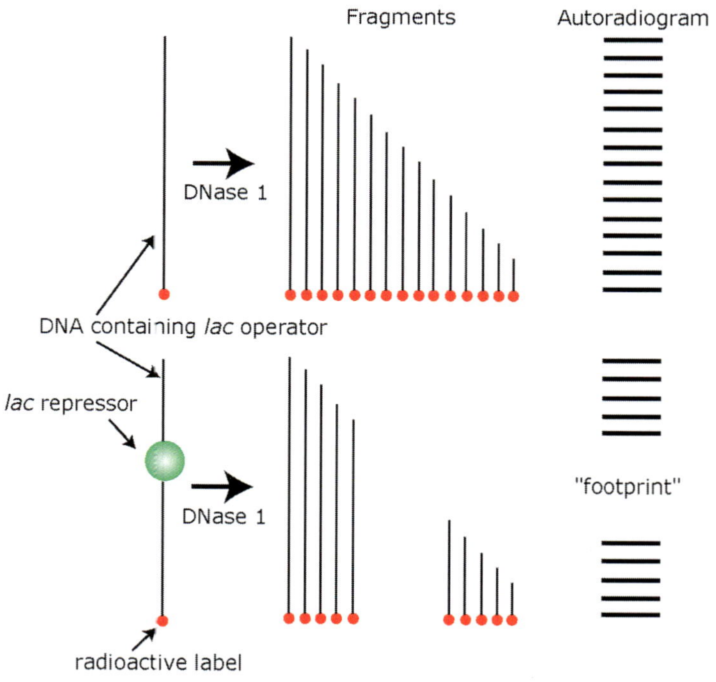

tion. The ability of an activator is essentially dictated by its homing DNA sequence; any gene that has the requisite recognition DNA sequence will be up-regulated by the activator molecule. Thus, for example, a yeast activator like GAL-4 can be recognized by its corresponding DNA sequence that exists in a desired mammalian gene, allowing up-regulation of the target gene. This promiscuity of the activators has been used advantageously in constructing various cloning vectors in gene therapy, and, probably most cleverly, in **two-hybrid systems** to identify protein–protein interactions. Alternatively, protein–protein interactions can also be identified using techniques like DNase I footprinting (see figure accompanying the topic ***Trans*-acting elements**) and electrophoretic mobility shift assay (EMSA).

The DNA-binding proteins can interact with each other via specific domains, resulting in their cooperative binding to DNA. Thus, the presence of one activator at a DNA sequence can recruit another activator (or repressor). The combined effect of multiple activators is shown to be synergistic. In some cases, the different activators may not physically interact with each other, yet they can bring about synergy of gene induction by simultaneously communicating with the transcriptional machinery. Besides affording the cell an ability to mix-and-match the activators for fine control of gene regulation, the presence of multiple activators may also be helpful in tackling the steric hindrance caused by nucleosomal proteins. The ability of a given activator to interact with different combinations of activator molecules in different cell types allows the cell type-specific regulation of genes. Such combinatorial control encourages evolution of regulatory networks.

Extracellular signaling molecules have been shown to play an important role in the action of the activators as well as repressors. Thus, the signaling molecules can liberate an activator from its inhibitory binding with a repressor in the cytoplasm, allowing it to be translocated to the nucleus (possibly through interaction with molecular chaperones), or they can trigger posttranslational modifications, such as phosphorylation, cleavage, or glycosylation, and convert it into an active form. Thus, pathways of signal transduction are intricately linked with those of gene transcription.

Suggested Reading

Kucharczuk, K.L. and D.J. Goldhamer. 1997. Nuclear DNA-binding proteins. Methods Cell Biol. *52*:439-472.

DNA METHYLATION

Linked Topics:
Bacteriophages
Genomic imprinting
Mismatch repair genes
Restriction endonucleases

Methylation is one of the most frequent postsynthetic modifications in DNA. Using the S-adenosyl methionine as the methyl group donor, the enzyme DNA methylase or DNA methyl transferase adds a methyl group to C5 of cytosine or N6 of adenosine. In prokaryotes, the restriction modification system consists of two components, namely, methyl transferase, which methylates indigenous DNA sequences, and restriction endonucleases, which cleave unmethylated DNA of the invading bacteriophages, thus protecting the host cell. Another role that DNA methylation plays in prokaryotes is the mechanism of DNA repair. Thus, the newly synthesized DNA strand during replication is distinguished from the parental DNA by delayed methylation of the new strand. Any mistakes in the newly replicated strand, therefore, are corrected in accordance with the parental DNA strand.

In eukaryotes, the base predominantly methylated is cytosine (resulting in formation of 5 methyl cytosine or 5mC) in CG dinucleotide (CpG). The methyl transferase methylates CG residues on the progeny strand paired with the hemimethylated parental strand. The levels of methylation are shown to be greatly variable across the organisms; *Drosophila* have very low C methylation, while vertebrates have the most CG residues methylated. Thus, while all vertebrate genes contain mCG, the invertebrate genes contain almost none.

Although the theory is controversial, DNA methylation is suggested to have a role in the regulation of transcription, because the presence of mCGs is shown to be related to the repression of transcription. This observation is substantiated by the fact that the inactive genes that are methylated can be reactivated by treatment with 5 azacytidine, which is an analog of C that prevents methylation. While gene activation by way of methylation loss remains to be proven unequivocally, the addition of methyl groups has been shown to suppress gene activation, as seen in X chromosome inactivation in female mammals. The mechanisms of the control of gene expression via methylation are possibly twofold: one, that it directly prevents transcription factors from binding to their recognition sites in the promoter, and two, the methylated C interacts with the other nuclear components to change the composition of chromatin, thus preventing transcription indirectly. The stretches of mammalian genome called **CpG islands** are known to be constitutively free of methylation. These are found at the 5′ ends of most vertebrate genes, particularly the housekeeping genes or tissue-specific expressed genes. These islands normally lack methylation, regardless of whether the gene is active or inactive (inactivated X chromosome is the only exception).

During the evolution of vertebrates, DNA methylation may have acquired other functions, such as genomic imprinting. Another important contribution that DNA methylation may have made to vertebrate development pertains, ironically, to the mutability of genomes. Normally, cytosine deaminates to give uracil, but methyl cytosine deaminates to yield thymine. The protein products of mismatch repair genes are rather inefficient in recognizing and correcting the TG mismatches, resulting in replacement of mCG by TG. Nearly a third of such point mutations in human genomes are causally related to genetic diseases. It might be the case, then, that while higher vertebrates have acquired finer controls over gene expression and parent-specific genomic imprinting of alleles through DNA methylation, this control probably has been traded for a higher rate of genetic diseases due to increased mutations.

Suggested Reading

Laird, P.W. 2003. The power and the promise of DNA methylation markers. Nat. Rev. Cancer *3*:253-266.

Fraga, M.F. and M. Esteller. 2002. DNA methylation: a profile of methods and applications. BioTechniques *33*:632-649.

DNA SEQUENCING

Linked Topics:

Gene cloning methodologies

Human Genome Project

Modifications of DNA sequencing

Pharmacogenomics and pharmacogenetics

Physical mapping

Sequence-tagged sites and content mapping

Single nucleotide polymorphisms

As the Human Genome Project has proceeded with increasing momentum through the past decade, the technology of DNA sequencing has come to be recognized as a pivotal tool in genomics. As an essential component of the ultimate step in physical mapping that provides the highest resolution, DNA sequencing has proven to be useful in mutation screening, clone verification, and sequence-tagged sites (STS) content and contig mapping after gene cloning. Besides genome mapping, it can also help predict the functions of certain polypeptides, provide information on population genetics, and help identify the gene mutations that are causally or genetically linked to human diseases. The latter aspect of mutation detection [or single nucleotide polymorphism (SNP) detection] is driving the fields of pharmacogenomics and pharmacogenetics and is expected to help screen people at high risk for disease development. Two traditional methods of DNA sequencing are Maxam-Gilbert's chemical cleavage and Sanger's dideoxy sequencing, while other modifications of DNA sequencing are described elsewhere.

Maxam-Gilbert's method (named after Allan Maxam and Walter Gilbert) involves chemical modification of purine and pyrimidine bases in a DNA sequence, end-labeled with radioisotopic [$^{32}P\gamma$]ATP and subsequent cleavage to leave a signature pattern of bands. Thus, in separate reaction tubes, the purines adenine and guanine are chemically modified by dimethyl sulfate to N_3-methyladenine and N_7-methylguanine, respectively; while both the pyrimidines, thymine, and cytosine are modified by hydrazine to 2-deoxyribosyl hydrazone. Further, on treatment with piperidine, the labeled DNA strands are cleaved at the 3′ position of sugar residues of the modified nucleotides, resulting in distinct base-specific fragment patterns. The concentration of the modifying and cleavage-inducing chemicals is adjusted so that the cleavage is random and partial, ensuring that the fragments are in overlapping lengths. The cleaved fragments from each reaction tube are separated by electrophoresis, according to the chain length, and identified by autoradiography. The nucleotide sequence is deduced from the position of bands in each of the four lanes in the gel. This technology has largely been abandoned due to chemical hazards and laborious nature.

Sanger's dideoxy chain termination method (named after Frederick Sanger) involves copying in different reaction tubes the same template single-stranded DNA molecules using a radioisotopically (or fluorescently) labeled primer and the arrest of the polymerization at any of the four bases by adding the corresponding dideoxyribonucleotide (ddNTP) derivative of the deoxyribonucleoside phosphates. Chain termination results because dideoxyribonucleotides lack 3′ hydroxyl (OH^-) groups essential for chain extension. Thus, for example, a tube containing an appropriate mixture of unmodified dATP and ddATP would result in incorporation of ddATP in place of dATP and random termination of synthesis at a location corresponding to dTTP in the template strand, and so on. It should be noted that the primer is extended 5′ to 3′ so that the terminated nucleotide is 3′, hence the sequence of the template strand is read beginning at the 5′ end. Thus, each tube contains a mixture of fragments that all terminate at the same nucleotide and differ in length by as little as one nucleotide. Until recent years, the fragments were separated by electrophoresis in 8% ultra-thin polyacrylamide gels and detected by autoradiography. Carrying out unlabeled reactions and subsequent silver staining of the gels has been another option. But the advent of fluorochromes with four distinct emission characteristics has permitted simultaneous chain extension and termination in the same tube and fragment separation in the same well of the gel. The products of all four reactions are detected by laser scanning at the fixed point across the base of the gel, fluorescence spectroscopy, and analysis by appropriate computer algorithms, finally resulting in a printout of the base sequence. This introduction of fluorescent sequence read-out has allowed easy automation, longer sequence reads per reaction, and tremendous increase in throughput. Sequencing RNA is also made possible by simply reverse transcribing the single-stranded RNA template while using ddNTPs to terminate primer extension as described above. Sequencing of ribosomal RNA across species has thus allowed growth of the field of **molecular systematics** for the study of evolution. Easy access to genomic DNA and cDNA sequences for numerous organisms via comprehensive bioinformatic databases has been a major development in biotechnology.

Suggested Reading

Marziali, A. and M. Akeson. 2001. New DNA sequencing methods. Ann. Rev. Biomed. Eng. *3*:195-223.

DNA TYPING: DNA FINGERPRINTING OR DNA PROFILING

Linked Topics:
Allele-specific oligonucleotides
Autoradiography
PCR and PCR optimization
Restriction endonuclease

Although the earlier biochemical identification techniques like blood group matching and isoenzyme patterning were useful for exclusion of an individual in genetic testing, they were not definitive for inclusion in cases where the genetic markers may be shared. The technique of DNA typing provides markers that are conclusive for both inclusion as well as exclusion.

DNA typing is the analysis of numerous hypervariable loci within a given genome to reveal an individual-specific unique DNA pattern. A hypervariable locus (minisatellite DNA) consists of a block of tandem repeats with a short core sequence. First discovered as polymorphic alleles in the D14s1 locus on human chromosome 14, many minisatellites have been reported in a number of human genes including insulin, myoglobin, and H-*ras* genes. Most of these have tandem sequences with 10–60-bp-long G-rich cores. Unrelated individuals have a variable number of such tandem repeats (VNTR), while closely related individuals share the homology of these repeats. The inter-individual variability is due to differences in either the length or the sequence of the repeats. Using the myoglobin gene repeat probes, Alec Jeffreys and his colleagues were the first to demonstrate the utility of VNTRs as unique genetic identification markers.

Each allelic haplotype is inherited in Mendelian fashion, making it possible to use these to determine family relationships and lineages. DNA fingerprinting is also used widely in ecological, anthropological, and evolutionary studies. Applications of DNA typing include establishing the identity of a murder or rape criminal (forensic science), determination of parentage (medico-legal cases and patent right establishment in plant seeds and cloned cell and/or animal strains), ascertaining reproductive success in males of wild animals and birds (population genetics), study of genetic relatedness (evolution), choosing most distant parents in an endangered species to improve breeding (eugenics), and others.

DNA extracted from various sources such as whole blood, blood clots, vaginal fluid, semen, hair roots, buccal cells, animal cells, and plant cells, can be subjected to (*i*) restriction endonuclease digestion by 4-bp cutters like *Hin*f1, *Alu*1, *Hae*III, or *Sau*IIIA, followed by restriction fragment-length polymorphism (RFLP); (*ii*) polymerase chain reaction (PCR) using locus-specific primers flanking the target repeats; (*iii*) direct DNA sequencing; or (*iv*) dot blot/reverse dot blot hybridization. In the latter case, either the target DNA or the allele-specific oligonucleotide probe is immobilized as dots onto a membrane. The final DNA pattern can be visualized by intercalating fluorescent dyes, autoradiography, or silver staining. The use of fluorescent dyes and development of laser-based detectors and computer software for data analysis has permitted automation of DNA profiling. Ability to infer genuine matches and nonmatches lies at the heart of successful of DNA typing. Although

DNA Typing

Tandem repeats in the myoglobin gene:

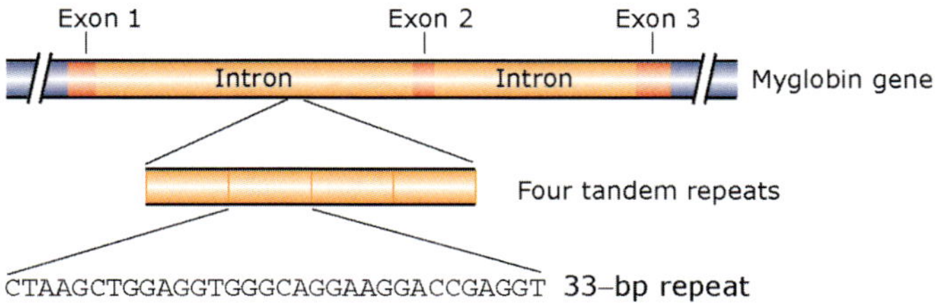

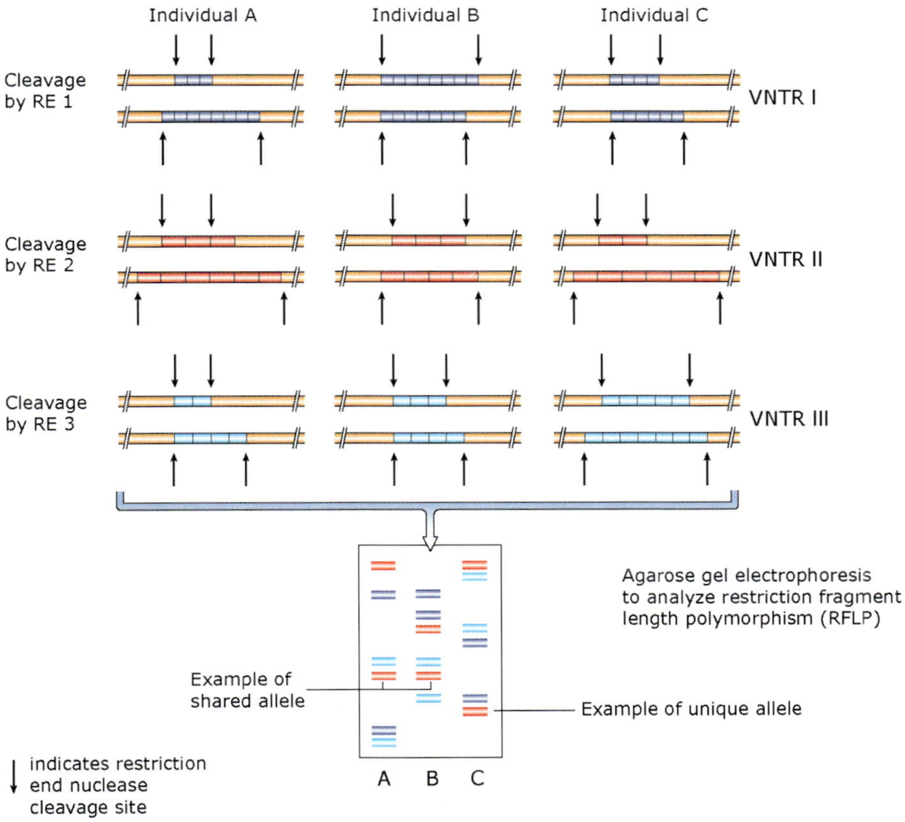

probability of two unrelated individuals sharing an identical allelic pattern is extremely low, the importance of preventing possible contamination of the prospective DNA sample is paramount.

Suggested Reading

Linacre, A. and D. Graham. 2002. Role of molecular diagnostics in forensic science. Expert Rev. Mol. Diagn. *2*:346-353.

EXON TRAPPING

Linked Topics:

Gene cloning methodologies

PCR and PCR optimization

Positional cloning (reverse genetics)

The various regularly used gene cloning methodologies rely on either expression of the genes or exploiting the propensity of many eukaryotic genes to be associated with CpG islands. However, these modalities are limited, either because of the restricted expression patterns of some genes or because many genes that are expressed in a tissue-specific or restricted manner do not have associated CpG islands. An alternative is therefore required to identify such genes by the ability of their exons to partake in an in vitro RNA splicing process. Exon trapping is an in vitro RNA splicing technique and is a component of the positional cloning (reverse genetics) approach.

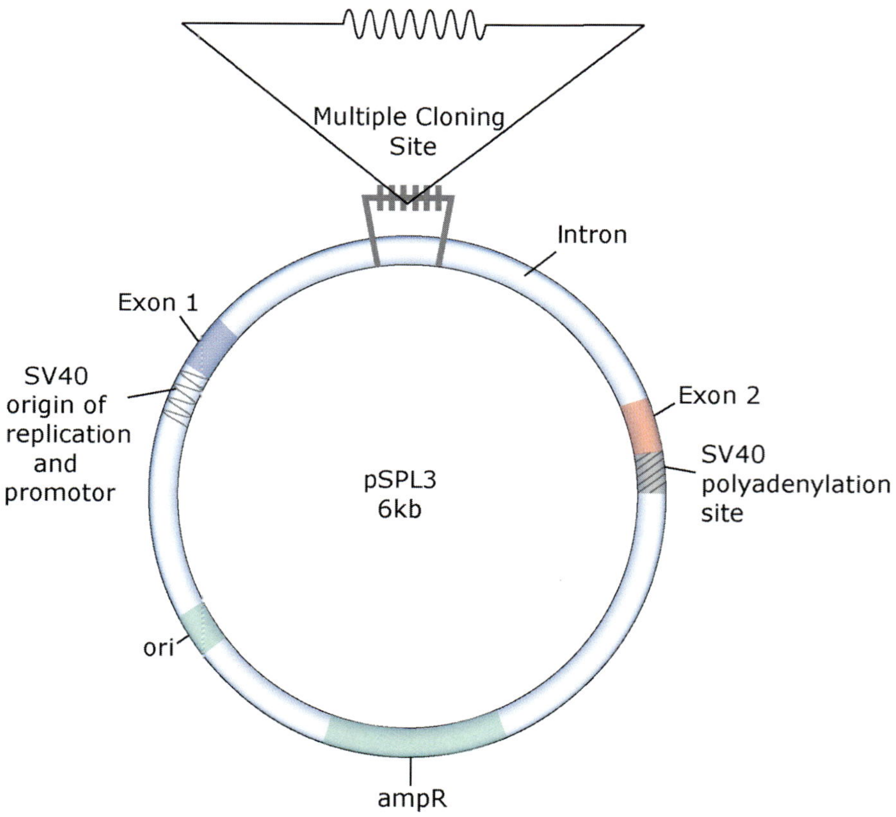

In vivo, RNA splicing involves fusion of exonic sequences at the RNA level and excision of intronic sequences. The enzyme complexes responsible for this fusion in vivo, **spliceosomes**, accomplish this by recognizing certain sequences at the exon–intron boundaries; a **splice donor** sequence at the junction between an exon and its downstream 3′ intron; and a **splice acceptor** sequence at the junction between the exon and its upstream 5′ intron. The splice donor site has a consensus sequence (C or

A) AG/GT (A or G) AGT, while the splice acceptor site has the consensus sequence (C or T)$_9$ N (C or T) AG/(G or A), where N can be any nucleotide, and the slash (/) indicates the exon–intron or intron–exon boundary for splice donor and splice acceptor, respectively. If a cosmid or another genomic cloning vector carries a foreign DNA segment with an internal exon flanked by intronic sequences, it will, in essence, contain functional splice donor and splice acceptor sites. Such trapped exons in cloned genomic DNA can be identified by subcloning the DNA into a suitable expression vector and transfecting it into an appropriate eukaryotic cell line wherein the insert DNA is transcribed into RNA and that undergoes splicing. Exon capture is followed by polymerase chain reaction (PCR), hence this technique is also known as **exon amplification.**

An example of a vector used in exon trapping is **pSPL3**. This vector contains three components of an artificial mini-gene: (*i*) a segment containing the simian virus 40 (SV40) origin of replication (ori) and a promoter sequence; (*ii*) two splicing component exons separated by an intron, which itself contains multiple cloning sites (MCSs); and (*iii*) an SV40 polyadenylation site. The recombinant DNA is transfected into a strain of monkey cells known as COS cells (CV1 cells containing defective origin of SV40 replication). The integrated SV40 dysfunctional sequence in COS cells allows vector DNA containing functional SV40 replication origin to replicate independently of cellular DNA. Transcription from the SV40 promoter results in an RNA transcript which, in absence of the foreign DNA insert containing exon, splices to include two exons of the mini-genes. If, however, the trapped foreign DNA includes another functional exon, then the foreign exon(s) can be spliced to the exons present in the vector's mini-gene. After reverse transcribing into cDNA and PCR amplification using a primer specific for vector exon sequences, one can detect the trapped exon by size fractionation in gels, which can lead to the recovery of the amplified exon from the genomic DNA.

Suggested Reading

Katoh, M. 2002. Paradigm shift in gene-finding method: from bench-top approach to desk-top approach. Int. J. Mol. Med. *10*:677-682.

Wapenaar, M.C. and J.T. Den Dunnen. 2001. Exon trapping. Application of a large-insert multiple-exon-trapping system. Methods Mol. Biol. *175*:201-215.

FLUORESCENCE RESONANCE ENERGY TRANSFER

Linked Topics:
Modifications of PCR
Real-time PCR
TaqMan® and molecular beacons

The most common application of fluorescence resonance energy transfer (FRET) principle is the labeling of real-time polymerase chain reaction (PCR). Here, if one end of a short oligonucleotide probe is synthesized to carry a fluorophore and the other a quencher, the fluorophore fails to fluoresce due to its proximity to the quencher. The quencher moiety [a nonfluorescent chromophore like 4-(4′-dimethylaminophenylazo) sulfonic acid (DABCYL)] receives the fluorescence resonance energy from the fluorophore (like fluorescein) and emits it as heat. The two routinely used real-time PCR applications are TaqMan® and molecular beacons.

Another emerging and exciting application of the FRET principle is in vitro and in vivo real-time tracking analysis of molecular interactions. Techniques of genetic engineering have been used to fuse cDNA for jellyfish green fluorescent protein (**GFP**) with that of any desired protein gene, and the resultant fusion protein then acquires green fluorescence property, which can be utilized to follow the transport and interactions of the protein in living cells. FRET with GFP detects the proximity of fluorescently labeled molecules over distances of 10–100 Å. Since this assessment requires spatial resolution that exceeds the optical limit of the light microscope, fluorescence digital imaging microscopy is used to visualize and follow the time course of the changes in FRET corresponding to cellular events at a millisecond time resolution. A digital imaging microscope such as a confocal microscope or a wide-field fluorescence microscope coupled to a charge-coupled device (CCD) camera can be used in these analyses.

A variety of GFP variants with spectral overlapping properties are now available. These can be employed to study homotypic (such as MDM2/MDM2) or heterotypic (such as p53/MDM2) protein interactions that are vital in the regulation of processes like cell cycle and apoptosis. Alternatively, spectrally overlapping GFP variants can be created as fusion products with the target protein, such that the interaction occurs at two different sites within the same protein, offering a unique opportunity to monitor the protein conformational changes in real time. The observation of such dynamic molecular events allows topological mapping of biomolecular interactions and is likely to provide crucial insights into biological pathways. Some of these data can also be translated into precise structural information using molecular modeling. The recent cloning of a sea anemone red fluorescent protein (RFP) has further expanded the spectrum of such analytical probes. Other FRET pairs include cyan fluorescent protein and yellow fluorescent protein, which are GFP mutants with altered spectral properties. These fluorescent proteins emit cyan and yellow light when exposed to UV radiation (302 nm).

Nuclear control of gene expression is another area that could benefit from this fascinating technology. Assembly of *trans*-activating factors in the nucleus and the complex behavior of key regulatory proteins can be examined in relation to chromatin. This assay can also be used to determine whether proteins that appear co-localized at the level of light microscopy interact with one another, and can be readily applied to samples prepared with standard immunofluorescence techniques using antibodies labeled with Cy™3 and Cy5 as the donor and acceptor fluorophores.

Suggested Reading

Eidne, K.A., K.M. Kroeger, and A.C. Hanyaloglu. 2002. Applications of novel resonance energy transfer techniques to study dynamic hormone receptor interactions in living cells. Trends Endocrinol. Metab. *13*:415-421.

Sekar, R.B. and A. Periasamy. 2003. Fluorescence resonance energy transfer (FRET) microscopy imaging of live cell protein localizations. J. Cell Biol. *160*:629-633.

Gaits, F. and K. Hahn. 2003. Shedding light on cell signaling: interpretation of FRET biosensors. Sci STKE *165*:PE3.

FUNCTIONAL GENOMICS

Linked Topics:

Aptamers

Microarrays

Mismatch repair genes

Pharmacogenetics and pharmacogenomics

Telomeres and telomerase

Two-hybrid system

While the term genomics refers to global analysis of structural elements of the entire genome of an organism, functional genomics is defined as an interpretation of the function of DNA sequence on a genomic scale. Functional genomics thus spans the gap between genomics and **proteomics** (a study of the entire protein complement of a cell or an organism). Functional genomics is assuming increasing importance, because knowing the structure of genes and the accompanying regulatory elements alone is inadequate for understanding the expression of these genes in the organism. Gene function results from the interaction of genomes within their macro- and microenvironment. Elucidation of such gene functions is achieved by large-scale characterization of the gene transcripts and their protein products. Functional genomics is expected to yield information about diverse biological questions, including not just how but why certain cells differ from the others in the body, how a given gene mutation leads to a particular lineage of tumor cell development, and how the pathogen–host protein circuitry lock into battle.

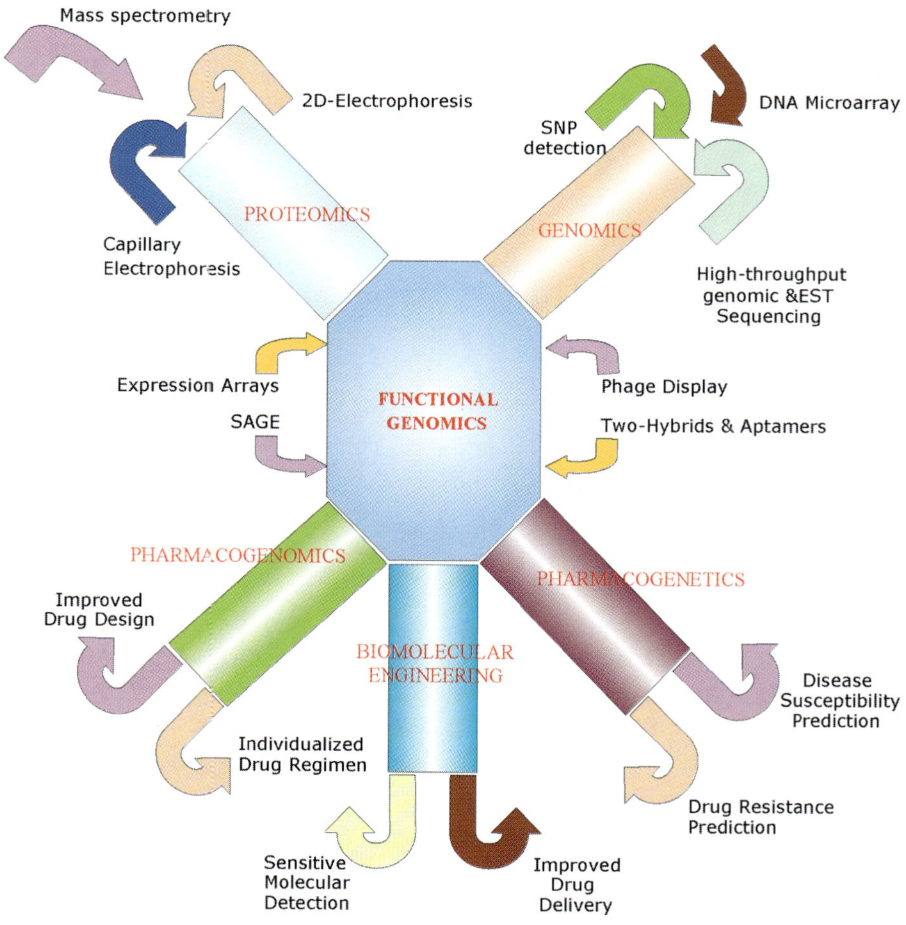

A protein involved in cell proliferation, stress response, and tissue homeostasis invariably functions via its interactions with other proteins and biomolecules, sometimes through multiple alternative shunt pathways intended to provide a backup. Applications of technologies such as the two-hybrid system and aptamers will permit identification of such associations and intermolecular communications. Other applications like microarrays, surface plasmon resonance, and 2-dimensional sodium dodecyl sulfate polyacrylamide gel electrophoresis (SDS-PAGE) will provide quantitative parameters of expression. Information about ligands that recognize protein products of all human genes (either antibodies or protein fragments) can be obtained via phage display technology. Such technologies, therefore, will constitute the tools-of-trade for functional genomics. High-throughput acquisition of knowledge of the expression of numerous genes in a qualitative as well as quantitative fashion is the cornerstone of the expanding fields of **expression profiling** and **expression signatures.**

One of the newer methods to obtain information about gene function is **comparative genomics.** Since evolution from common ancestors is known to preserve gene structures and functions across the development of new species, studying gene functions in one organism can provide valuable information about the function of related genes in others. A number of genes have indeed been isolated from a given species and their functions understood, based on knowledge about related genes in another species. Some of the most celebrated genes identified through universality of gene structure and biologic mechanisms of function are the human telomerase gene *hTERT* (which was tracked down due to its sequence and functional homology with yeast and tetrahymena genes) and the human mismatch repair genes (which have homology with *Escherichia coli* and yeast genes).

Creating a database of functional mechanisms of proteins is expected to provide a strategic rationale for efficient therapeutic intervention. This, in turn, will lead to identification of potential drug targets, which is the foundation of pharmacogenetics and pharmacogenomics.

Suggested Reading

Chanda, S.K. and J.S. Caldwell. 2003. Fulfilling the promise: drug discovery in the post-genomic era. Drug Disc. Today *8*:168-174.

GEL-BASED SNP DETECTION TECHNOLOGIES

Linked Topics:
Autoradiography
DNA sequencing
Mismatch repair system
PCR and PCR optimization

Gel-based single nucleotide polymorphism (SNP) detection technologies are based on the distinction between mutant and normal genes through amplification of their structural (conformational) differences, either at the single- or double-stranded DNA level. Typical polyacrylamide matrix (or modified forms of polyacrylamide with improved ability to discriminate between mutant and normal alleles) is employed in these assays. While analysis of a change of one or more nucleotides in single-stranded DNA molecules uses the altered folding patterns of these strands [as in single-strand specific conformation polymorphism (SSCP), see below], analysis of such changes for double-stranded DNA relies on formation of a DNA heteroduplex between normal and mutant DNA strands, where a conformationally distinct heteroduplex runs slower than a homoduplex during electrophoresis in a polyacrylamide gel. These mobility differences can be amplified by altering gel parameters such as pH, temperature, and matrix formulation, leading to enhanced resolution. The slower migrating duplex can be visualized in a gel either by autoradiography or fluorescence detection techniques. The results are often confirmed through direct DNA sequencing.

SSCP, the most widely used technique for single-stranded DNA molecules, is an open SNP detection system, in that it can detect both known and unknown polymorphisms in genes. Typically, polymerase chain reaction (PCR) products smaller than 300 bp are denatured and allowed to fold back on themselves. Any conformational changes are identified by resolution in nondenaturing polyacrylamide gel electrophoresis (ND-PAGE) over variable ionic and temperature conditions (90–130 mM Tris-buffered EDTA, 10°–20°C). However, since not every SNP results in a conformation change, lack of SSCP may not mean absence of a mutation.

In **heteroduplex analysis** (**HA**), higher temperature or denaturant concentration in a resolving gel causes denaturation around the region of mismatch, leading to an abrupt decrease in the mobility of the heteroduplex. A GC clamp is introduced at one end of the heteroduplex by PCR primers, which avoids complete separation of the two strands. Depending on whether change in denaturant concentration or temperature is used to form a denaturation gradient across the gel surface, either denaturing gradient gel electrophoresis (**DGGE**) or temperature gradient gel electrophoresis (**TGGE**) technique is used in HA. Denaturing high performance liquid chromatography and denaturing capillary electrophoresis based upon the same principles provide much higher resolution but are more expensive.

Although useful for identifying SNPs over a 100- and 400-nucleotide-long DNA, SSCP and HA do not provide SNP location. **Enzyme mismatch cleavage** (**EMC**) or **chemical mismatch cleavage** (**CMC**) at mismatch sites in a heteroduplex permit SNP identification even in relatively long DNA fragments up to 2 kb. EMC uses the property of bacteriophage enzymes T4 endonuclease VII and T7 endonuclease I (resolvases) to cleave the site of mismatch, while CMC uses osmium tetroxide and

Gel-Based SNP Detection

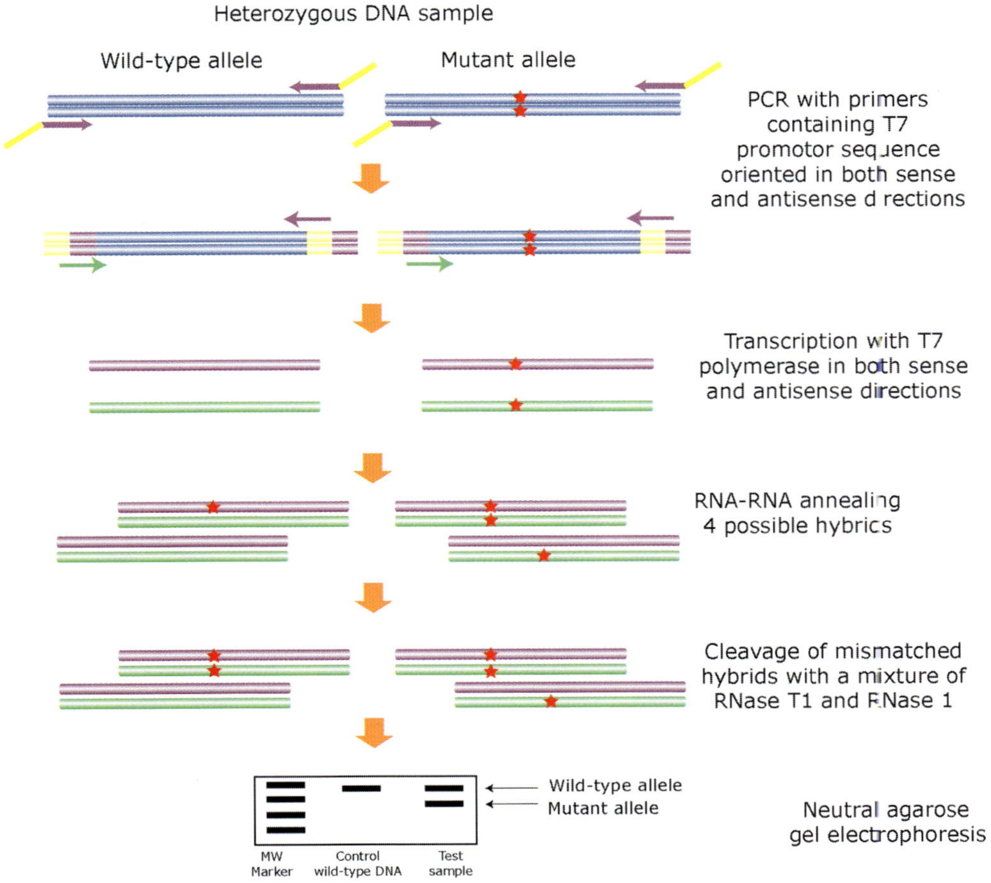

hydroxylamine to chemically alter mismatched bases, and piperidine is used to cleave the modified base. The mismatch repair system in *Escherichia coli* has provided a number of enzymatic reagents for detection and localization of mutations. Advances due to fluorescent PCR primers have reduced the hazards associated with radioisotopic ^{32}P-labeling and the identification of such SNPs. **Bidirectional dideoxy fingerprinting (Bi-ddF)** is the technique in which dideoxy chain termination (used in DNA sequencing) is combined with SSCP to detect and localize SNPs. Here, sense and antisense primers are extended with dNTP mixtures, containing either ddATP or ddCTP, and the products are resolved through a nondenaturing gel. An anomalous band with respect to control DNA denotes the SNP position.

Suggested Reading

Andersen, P.S., C. Jespersgaard, J. Vuust, M. Christiansen, and L.A. Larsen. 2003. Capillary electrophoresis-based single strand DNA conformation analysis in high-throughput mutation screening. Hum. Mutat. *21*:455-465.

Nataraj, A.J., I. Olivos-Glander, N. Kusukawa, and W.E. Highsmith, Jr. 1999. Single-strand conformation polymorphism and heteroduplex analysis for gel-based mutation detection. Electrophoresis *20*:1177-1185.

Linked Topics:

Expressed sequence tags

Genomic DNA or cDNA libraries

mRNA differential display

PCR and PCR optimization

Physical mapping

Positional cloning

Southern and Northern hybridization

Subtractive hybridization

Vectors

Yeast artificial chromosome libraries

cDNA enrichment cloning: Most initial attempts at gene cloning employed mRNA as a starting point. While genomic DNA from different cells pimarily consists of a collection of the same sequence complements, the differences lie in mRNA populations. As a result, cDNA libraries from different cell types can identify tissue-specific cDNA clones. Polymerase chain reaction (PCR)-based mRNA differential display method (**DD-RT-PCR**) can selectively amplify cDNAs corresponding to subsets of transcriptionally active genes and is an advancement that permits identification and cloning of unique sequences.

Location-based cloning: Physical mapping and positional cloning involve attempts to identify gene clones at specific subchromosomal locations. Thus, for example, the subchromosomal gene cloning strategy involves physical microdissection of a single chromosomal band from a cytogenetic preparation using ultrafine needles or laser beams. The DNA extracted from the excised band is used to construct DNA libraries. The construction of these microdissection libraries, however, is cumbersome because of contamination by extraneous DNA and is largely superseded by the development of yeast artificial chromosome (YAC) libraries. Another variation is subtraction cloning, which uses selection of unique DNA sequences by subtractive hybridization. The dystrophin gene was cloned in such a way. This technique tends to work better with cDNA rather than with genomic DNA. An advanced version of subtractive hybridization is representational difference analysis (RDA).

Protein-based cloning: This strategy uses the ability to isolate and purify a specific intracellular or secreted protein from a target cell and generating specific antibodies by injecting the purified protein preparation into rabbits or mice. The total cDNA is cloned in expression vectors to yield an expression cDNA library, which is then screened for a positive clone using the antibodies to the target protein. Alternatively, a partial amino acid sequence is obtained for a reasonably purified protein, and the corresponding cDNA is amplified by using degenerate PCR. This cDNA can then be used as a Southern hybridization probe for isolating the corresponding genomic DNA clone from a genomic library.

Homology-based cloning: Genes of an organism often show sequence homology with those in the same or other evolutionarily conserved species. A DNA sequence from a given species can, therefore, be used to screen for a similar gene from other related organisms. This can be done through molecular hybridization (by using a labeled DNA clone to screen a genomic or cDNA library by Southern hybridization, which is the basis of **zoo blot** hybridization) or by means of PCR (using degenerate PCR primers designed to correspond with short conserved amino acid consensus sequences). A more common method in recent years, however, is the use of expressed sequence tags (ESTs).

Gene cloning by computer analysis of the DNA sequences: As the Human Genome Project for complete DNA sequencing nears conclusion, the nucleotide sequence for numerous genes and inferred amino acid sequences are being recorded in public and corporate electronic databases. Computer analysis uses conserved exon–intron junction homology of related genes to predict open reading frames (ORFs) and exons in an uncharacterized genomic sequence, and the novel gene thus identified is then cloned. The most notable recent example of such cloning is the human telomerase (telomere repeat transferase, *hTERT*) gene, which used a combination of computer analysis of sequence homology between yeast and human genes and EST procedures.

Suggested Reading

MacNeil, D.J. and D.H. Weinberg. 2000. Homology-based cloning methods. Identification of the NPY Y2, Y4, and Y6 receptors. Methods Mol. Biol. *153*:61-70.

GENE TARGETING

Linked Topics:
PCR and PCR optimization

Delivery of exogenous DNA sequences into target cells forms the basis of the exciting science of gene therapy. Methods such as micro-injection, electroporation, vector-mediated transfection, liposome-mediated transfer, and gene targeting, can be used to effect such a delivery.

Use of homologous recombination to alter specific sequences in endogenous genes is referred to as gene targeting. This is a form of in vivo mutagenesis, whereby the sequence of a particular gene is selectively modified within an intact cell. It can, therefore, be viewed as an artificial site-directed mutagenesis in vivo.

If gene targeting results in the knock-out or null mutation, it can inactivate expression of the gene. Loss of function of a specific gene, conversely, can be used to test where and when in development that gene is required. The same approach can be potentially used to restore a normal phenotype by replacing a mutant gene sequence by a correct one. The latter promises to have therapeutic value in treatment of genetic diseases. The structure–function relationship of specific codons and amino acids can also be examined using this method. Another area where gene targeting can be employed is in the assessment of tolerance of structural genetic changes, while still retaining function of a protein product of a gene.

Although gene targeting in mammalian somatic cells, like myoblasts, has been attempted successfully, mouse embryonic stem (ES) cells are by far the preferred targets for such a genetic manipulation. After ex vivo engineering by gene targeting, the ES cells are introduced into the blastocyst of the foster mother, and the progeny mouse is obtained with the desired genetic change in all its nucleated cells.

Since homologous recombination in mammalian cells is a rare event, its frequency is increased by using isogenic mouse strains so that there is a high degree of sequence homology. In spite of this, identifying desired recombination events against the background of random integration events is very difficult. Hence, a targeting vector with a suitable selectable marker gene (e.g., gene for neomycin resistance) is employed to introduce the foreign DNA sequence. The method of choice for introduction of such a vector into the ES cells is electroporation. The confirmation of a successful homologous recombination event can be further augmented by polymerase chain reaction (PCR) using primers for the selectable marker gene and the sequences within the host chromosomal gene of interest.

Two types of targeting vectors have been used: insertion vectors and replacement vectors. Insertion vectors introduce the entire sequence into the recipient chromosomal locus by a single reciprocal recombination. This is a relatively safer and more sought after method to bring about knock-out mutations. Replacement vectors enable substitution of some of the chromosomal sequences by only a requisite part of the vector sequences by double reciprocal recombination events. Development of targeting vectors is continually being upgraded, and gene targeting may form a major tool in the armory of gene-based therapeutics for genetic diseases.

Insertion vector: Single site-specific recombination event

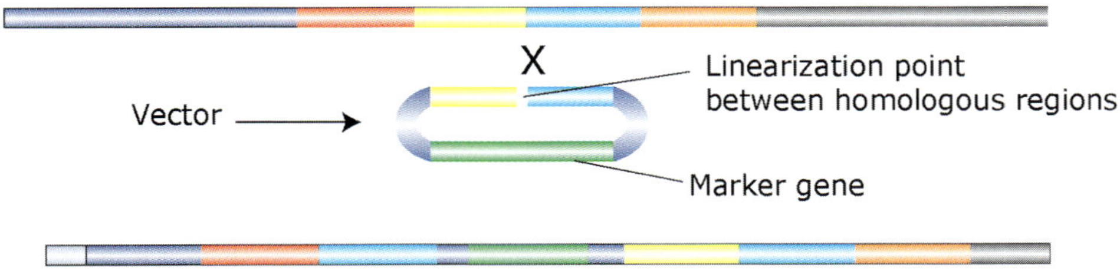

Resultant insertion of entire vector, in reversed orientation

Replacement vector: Double site-specific recombination events

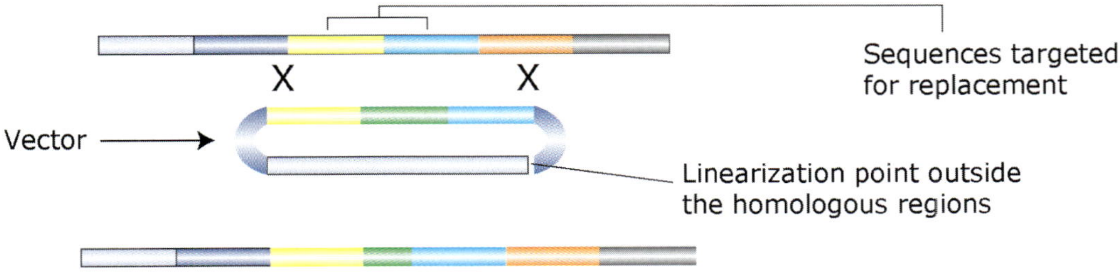

Resultant replacement of sequences from the host genome by vector sequences, in orientation same as the vector

Suggested Reading

Melton, D.W. 2002. Gene-targeting strategies. Methods Mol. Biol. *180*:151-173.

Kuhn, R. and R.M. Torres. 2002. Cre/loxP recombination system and gene targeting. Methods Mol. Biol. *180*:175-204.

DeChiara, T.M. 2001. Gene targeting in ES cells. Methods Mol. Biol. *158*:19-45.

GENE THERAPY

Linked Topics:
Antisense therapeutics
Genetic therapy
Pharmacogenomics and pharmacogenetics
Plasmids

While gene therapy has been the most publicized modern advancement in medicine, it really forms just one component of genetic therapies—the other components being antisense therapeutics and pharmacogenomics.

The classical definition of gene therapy includes the transfer of "healthy" genes into cells of an organism to replace or restore the function of "defective" genes and cure the disease. As the public concern over ethical aspects continues to be addressed by scientists, three very diverse gene transfer technologies are being developed simultaneously: development of synthetic vectors, improvement of virus-derived vectors (including retroviral, lentiviral, adenoviral, adenoassociated viral, and herpes viral vectors, all of which can be used as Trojan horses to transport a therapeutic gene into the target cell), and gene-altered stem cells. The vectors based on plasmids are simple and immunogenic but have very low efficiency of gene transfer and reduced stability. Compared to synthetic vectors, viral vectors have higher infectivity and better persistence, but not in human cells. Since they are modified ex vivo, gene-altered stem cells are safer than plasmid DNA and viral vectors. However, the downside is that not all cell types can be manipulated ex vivo. The current disease targets for gene therapy include genetic diseases, cancer, cardiovascular diseases, and infectious diseases such as acquired immunodeficiency syndrome (AIDS). The number of clinical trials in gene therapy continues to grow. As of July 2002, there were a total of 600 completed, ongoing, or pending trials worldwide, with the majority in the United States. Only about 13% of the protocols address genetic diseases, and even smaller proportions are treating AIDS and cardiovascular diseases. More than 63% of protocols deal with cancer. For cancer, gene therapy can be achieved by either introducing a tumor suppressor gene or blocking an oncogene function. Alternatively, the antitumor immune response can be stimulated.

Therapeutic DNA transferred via plasmid DNA transfection or electroporation, or both, has been shown to elicit an immune response. Such **DNA vaccines** are potentially useful in mice and rabbits but have not yet been tested in primates. **Chimeraplasty** is a creative alternative for gene delivery-based therapy, which aims at permanent repair of the patient's own genes so as to restore an aberrant gene function, instead of replacing or modifying the genes. This technique uses a single-stranded DNA/RNA chimera. The chimera is a self-complementary stretch of RNA (containing sequence that does not need repair) bracketing a small self-complementary stretch of DNA (containing the correction sequence). When introduced into cells, the mismatched bases between the chimera and the genic DNA are recognized by enzymes like mutS, which are products of mismatch repair genes. The mismatches in the target DNA are corrected, bringing about permanent alteration in the genomic DNA sequence. Since RNA/DNA hybrids form more readily than DNA/DNA, the flanking RNA forms a clamp to allow targeting of the molecule, while the DNA stretch effects the mismatch repair. Notably, the resulting organism need not be considered genetically modified, because foreign DNA is not introduced.

Chimeraplasty

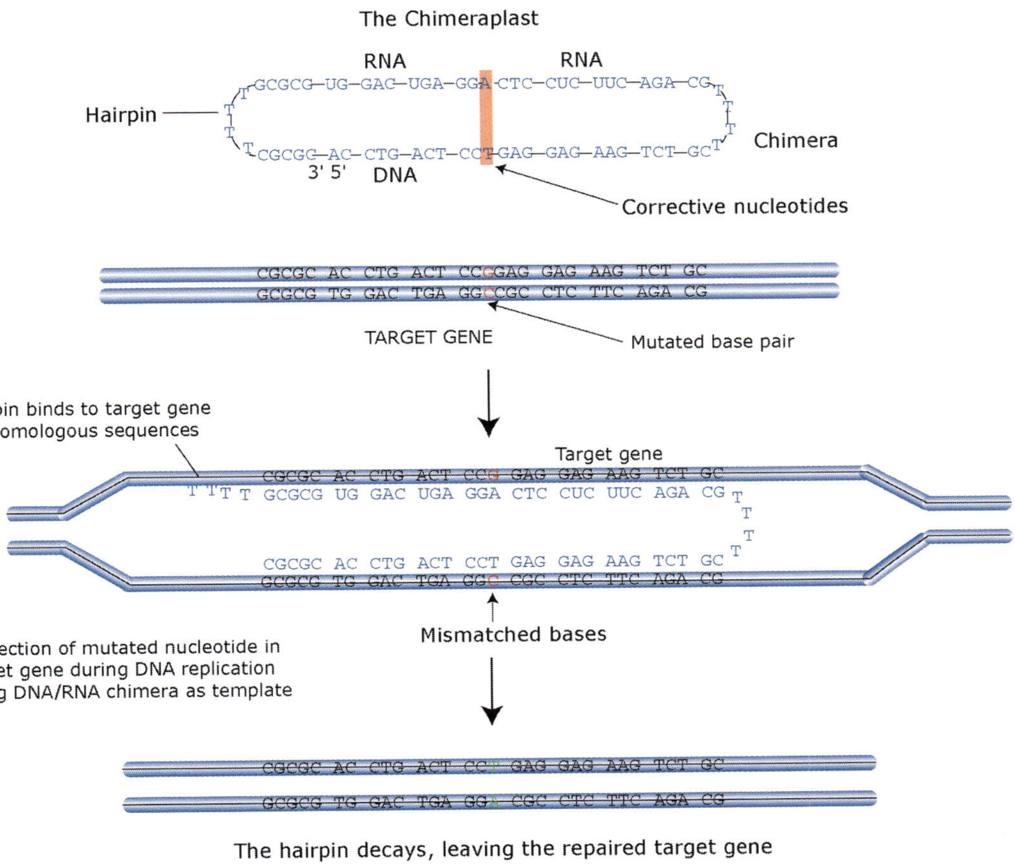

Modulation of gene therapy: Depending upon the target of treatment (ex vivo cells, in vivo tissues, or the circulatory system) or upon the treatment indication (short-term, long-term, or lifetime), the gene therapy dose regimen is variable. Also, the contribution by the immune system may be encouraged (for cancer) or suppressed (in other diseases). Future improvement of gene therapy needs engineering vectors with better efficacy at smaller doses, reaching prolonged therapeutic expression levels and the potential for large-scale manufacture of simple and synthetic vectors.

Suggested Reading

Datar, R.H. 2001. Nucleic acids in therapy. Nat. Med. J. India *14*: 290-296.

Noguchi, P. 2003. Risks and benefits of gene therapy. N. Engl. J. Med. *348*:193-194.

GENETIC TESTING

Linked Topics:
Human Genome Project

Genetic screening or genetic testing can be defined as testing on a population basis to identify a subset of individuals at high risk of developing a specific disorder or of transmitting a specific disorder to the next generation. Selected single gene diseases can be amenable to prenatal presymptomatic diagnosis by DNA analysis. These can be roughly grouped into three categories: autosomal dominant diseases, autosomal recessive diseases, and X-linked recessive diseases. Examples of the autosomal dominant diseases are myotonic dystrophy, Huntington's disease, familial breast cancer, adult polycystic kidney disease, and neurofibromatosis. The autosomal recessive diseases include sickle cell anemia, thalassemia, cystic fibrosis, phenylketonuria, and Tay-Sach's disease, among others. Some examples of X-linked recessive disorders are hemophilia A and B, Duchenne's and Becker's muscular dystrophy, fragile X syndrome, and ornithine transcarbamoylase deficiency.

Before undertaking any genetic testing, it must be ensured that the test has high sensitivity, high specificity, a low false-positive rate, and low false-negative rate. The sensitivity of a test is defined as the frequency with which the test yields a positive result when the disease is present. The specificity is the frequency with which the test is negative when the disease is absent. The false-positive rate is the proportion of all positive test results occurring in individuals who are not affected. The false-negative rate is the proportion of all the negative results seen among individuals affected with the disease. The sensitivity reflects the ability of the test to detect affected individuals, whereas specificity indicates the ability of the test to exclude unaffected individuals.

Identification of genes that are proven to account for susceptibility to serious and potentially preventable diseases, such as cardiovascular disease and cancer, supports the need for genetic screening programs to identify individuals at risk. Genetic testing for mutations in *BRCA1* and *BRCA2* genes is a good example. Mutations in these genes account for the majority of the breast cancers in families with four or more affected close relatives with breast cancer and an even higher proportion of those families with both breast and ovarian cancer. Many different mutations have now been identified in these two genes, especially in the Ashkenazi Jewish population.

Several major problems will have to be solved before generalized genetic screening programs can be undertaken. The issues that need to be sorted out will be the same ones that are being addressed by the ethical, legal, and social implications (ELSI) initiative of the Human Genome Project. An important aspect of genetic screening concerns the patient's right to decide the future course after the outcome of the test is made available.

The genetic screening programs already undertaken include those for cystic fibrosis in northern European populations, β-thalassemia in Sardinia and Italy, and Tay-Sach's disease in the Ashkenazi Jewish population. They have all resulted in reduction of the frequency of these diseases. Thus, the genetic screening approaches to identify high-risk individuals in a population will hopefully, in the long run, result in the reduction of overall risk for a given disease.

Suggested Reading

Grody, W.W. 2003. Molecular genetic risk screening. Ann. Rev. Med. *54*:473-490.

Khoury, M.J., L.L. McCabe, and E.R. McCabe. 2003. Population screening in the age of genomic medicine. N. Engl. J. Med. *348*:50-58.

GENETIC THERAPY

Linked Topics:

Antisense oligonucleotides

Aptamers

DNA sequencing

Gene cloning methodologies

Gene therapy

Human Genome Project

Modified oligonucleotides

PCR and PCR optimization

Pharmacogenomics and pharmacogenetics

Ribozymes

Three technologies based on genetic information comprise genetic therapy: **gene therapy, antisense therapeutics,** and **pharmacogenomics/ pharmacogenetics**. Historically, biomedical research has seen a transition of the quest through various stages. Starting with studies aimed at finding single gene–single function associations, the inquiries proceeded to genetic maps with a catalog of genes, to a holistic gene complement catalog for an organism, and finally, to using the interrelationships of multiple genes and their functions to produce a molecular profile of an organism. This information has the potential to dramatically improve human health; so while these studies progressed, the three aspects of genetic therapy evolved, initially independently, and then gradually merging with significant overlaps.

There is a growing debate about the ethical, legal, and social implications (**ELSI**) that need to be addressed before these technologies are used as routine clinical alternatives or to supplement the existing treatment modalities. Recognizing that the information gained from the Human Genome Project can have far-reaching implications for individuals, families, and society, the organizers of the Project, starting in 1989, created a mechanism for public debate to develop a consensus for the interpretation and use of the genetic information. The ELSI initiative will also decide ways to permit access to the information and to protect individuals from the harm that might result from its improper disclosure or use.

Gene therapy has contributed greatly to the developments in recombinant DNA techniques, such as human gene cloning, DNA sequencing, polymerase chain reaction (PCR), gene delivery, and others, and attempts are increasingly being made to replace a defective gene or transfer a missing function gene into human cells ex vivo or in vivo. The ex vivo gene transfer experiments use viral delivery, liposome-mediated transfer, or electroporation. Many clinical trials are in progress that employ different methods of gene delivery. The route of delivery in vivo could be systemic injections (for localized malignant tumors), local infusions (for prevention of arterial restenosis), or micro-injection of recombinant viral vectors by stereotactic surgery (as for brain tumors). Microparticle-coated gene bombardment using a gene gun is a technique that is being developed in animal models for eventual use in humans. The latest efforts employ the chimeraplasty technique for correcting the gene mutations in situ, which may have an important role in future genetic therapy.

Antisense therapeutics, which takes recourse to the sequence-specific antisense binding of nucleic acids to either complementary molecules (as in case of antisense oligonucleotides for DNA as well as RNA or ribozymes for RNA) or to target proteins (as in aptamers), is a novel branch of genetic medicine that has been in development. Several clinical trials are under way to assess the efficacy of this approach. Accessory technologies, such as selection by exponential amplification (**SELEX**), for aptamer production and synthesis of modified oligonucleotides for antisense oligonucleotide drugs are pivotal to the progress of antisense therapeutics. Since sequence-independent interactions of antisense molecules with nontarget biomolecules are known to occur, caution in human application is necessary.

Pharmacogenomics and pharmacogenetics provide advances in the analysis of molecular genetics, which increasingly result in extensive information about the molecular etiology of diseases and the identity of the genetic makeup of pathogens. Both these data sets are being used to tailor rational drugs for individual patient-specific treatment.

Suggested Reading

Kurreck, J. 2003. Antisense technologies. Improvement through novel chemical modifications. Eur. J. Biochem. *270*:1628-1544.

Opalinska, J.B. and A.M. Gewirtz. 2002. Nucleic-acid therapeutics: basic principles and recent applications. Nat. Rev. Drug Disc. *1*:503-514.

Knauert, M.P. and P.M. Glazer. 2001. Triplex forming oligonucleotides: sequence-specific tools for gene targeting. Hum. Mol. Genet. *10*: 2243-2251.

GENOMIC DNA AND cDNA LIBRARIES

Linked Topics:

Bacterial artificial chromosomes

Bacteriophages

Human Genome Project

P1-derived artificial chromosomes

PCR and PCR optimization

Plasmids

Positional cloning

Restriction endonucleases

Southern and Northern blotting

Yeast artificial chromosomes

Cell-based cloning of DNA, unlike polymerase chain reaction (PCR), permits very large amounts of a DNA clone. It offers a wide range in the size of DNA fragments that can be cloned and permits rapid subcloning into useful vectors engineered to allow expression of the cloned gene in cells or even whole animals. Although PCR allows rapid cloning of an initial smaller gene fragment (except for inverse PCR), cell-based cloning is often the preferred mode of construction for genomic and cDNA libraries, which are then screened by PCR to identify and isolate clones of cells containing the gene sequence of interest.

Modern day cloning approaches offer the possibility of making comprehensive collections of DNA clones from extremely complex starting DNA populations. Such collections are termed DNA libraries. This approach enables DNA sequences that are very rare in the starting population to be presented in a library of DNA clones. They can then be isolated individually, by selecting a suitable host cell colony and amplifying it. Two basic variations of this method have been employed, genomic DNA libraries and cDNA libraries. While vectors like plasmids and λ bacteriophage derivatives are preferred for constructing cDNA libraries, the genomic DNA libraries require vectors such as cosmids, bacteriophages, bacterial artificial chromosomes (BACs), P1-derived artificial chromosomes (PACs), or yeast artificial chromosomes (YACs) to accommodate larger inserts.

Construction of **genomic DNA libraries** typically involves digestion of genomic DNA with a restriction endonuclease cleaving at a 4-bp recognition sequence. Theoretically, this sequence will occur once every 275 bp on average, and there will be very few DNA sequences lacking the recognition site. Partial restriction digestion is favored, allowing cleavage at only a few of the potential restriction sites. This random fragmentation results in production of large yet overlapping fragments for insertion into cloning vectors, thus ensuring representation of most of the starting DNA in the library. The complexity of a genomic DNA library is defined by the number of independent DNA clones included; the higher the number, the greater is the complexity. The ability to fractionate individual human chromosomes or even specific chromosome bands has resulted in construction of chromosome-specific libraries and chromosome band-specific libraries. Such genomic DNA libraries are a valuable and integral resource for Human Genome Project.

The functional constraint on genomic clones due to inability of prokaryotic host systems to remove the introns is circumvented by synthesizing complementary DNA (cDNA) from a large number of functional eukaryotic mRNA and cloning the cDNA, which results in construction of a **cDNA library**. Since gene expression in different cells is often differentiation- and lineage-specific, as well as development-specific, the starting material is usually mRNA from specific tissue or specific developmental stages during embryogenesis, which is converted into cDNA using reverse transcriptase. The cDNA library can be used for a number of purposes. The cDNA itself can be used for determination of its entire DNA sequence, prediction of its corresponding amino acid sequence from the knowledge of the genetic code, or as a hybridization probe to identify related DNA or RNA sequences using Southern or Northern blotting. In addition, the cDNA can be used as a probe to clone the complete structural gene, which includes all the exons, introns, and the adjacent promoter sequences. Positional cloning has made it possible to clone a hitherto unknown gene based solely on genetic information, with no knowledge of the corresponding protein. The relevant cloned cDNA can be used to study the function and possible role of the newly discovered gene product in the corresponding human disease.

Use of Cosmids in Construction of Genomic DNA Libraries

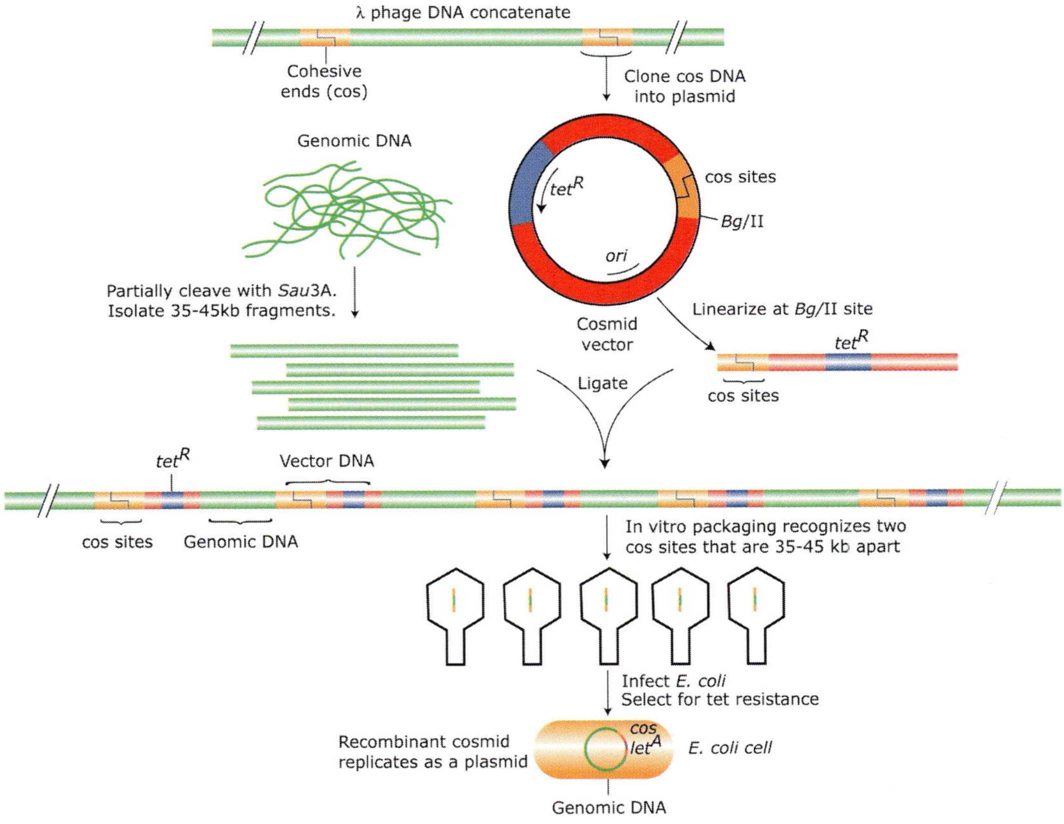

Suggested Reading

Ying, S.Y. 2003. Complementary DNA libraries: an overview. Methods Mol. Biol. *221*:1-12.

Zhu, J. 2002. Use of PCR in library screening. An overview. Methods Mol. Biol. *192*:353-358.

Campbell, T.N. and F.Y. Choy. Approaches to library screening. J. Mol. Microbiol. Biotechnol. *4*:551-554.

GENOMIC IMPRINTING

Linked Topics:
DNA methylation

Mendelian genetics assumes that both the paternal and the maternal alleles of biallelic genes are expressed unless one or both copies have sustained mutations affecting their expression. But this does not hold true for several biallelic genes known in humans and other mammals where, at least in some cells, only one parental allele is normally expressed—either the paternal or the maternal. This occurs although both the parental alleles are quite capable of normal gene expression. If two copies of a chromosome with a nonexpressed allele are inherited, gene function will be abolished by an epigenetic mechanism, even if both gene copies themselves are genetically capable of functioning. On the other hand, deletion (or mutational inactivation) of an active allele of a gene would result in structural monosomy, but functional nullisomy, for the region. The result is a functional hemizygosity, such that only a half of the maximum gene product is obtained. Genomic imprinting is one of the mechanisms accountable for such a mono-allelic expression. (Apparently, most genes are not subject to imprinting, or else simple Mendelian characters would not have been so numerous.)

The Prader-Willi syndrome is related to microdeletions in the chromosomal region 15q11-q13, which is inherited from the father. Interestingly, Angelmann syndrome patients present with the same microdeletions on chromosome 15 as in Prader-Willi syndrome patients, but the deletions are always on the chromosome inherited from the mother. Beckwith-Wiedemann syndrome also occurs in children who inherit the gene from their mother. Recent studies have demonstrated high incidence of cancer in infancy and early childhood for Beckwith-Wiedemann syndrome patients.

Examples of allelic loss in many tumors involving only paternal alleles are also known. The diploid hydatidiform moles are the human placental tumors with two paternally derived sets of chromosomes, but none maternal. Ovarian teratomas, on the other hand, are benign tumors containing cells bearing only maternally derived chromosomes.

These pathological conditions illustrate that the absence of alleles from one of the parents, resulting in abnormal development, is due to an imprint of their parental origin borne by their genomes. This imprint ought to be reversible, since a man passes on genes with his own paternal imprint even if the genes had come to him with a maternal imprint via his mother. This suggests that some mechanisms must be able to distinguish between maternal and paternal alleles, and as chromosomes pass through the male and female gametogenesis, they must acquire some sort of imprint to signal a difference between paternal and maternal alleles in the developing organism. Imprinting seems to operate at the transcriptional level, and the physical basis of imprinting probably involves a process of DNA methylation. Imprinting may vary between tissues and during development. Perhaps, the process of genomic imprinting allows an extra level of control of gene expression.

The biological purpose of imprinting is not yet fully understood. A tantalizing but unsubstantiated theory implies that imprinting is a result of evolutionary conflict between two selfish genomes in placental mammals. This derives from the fact that paternal genes are preferentially expressed in trophoblasts and extra-embryonic membranes, while maternal genes are better expressed in the embryo, suggesting that the paternal genome in an embryo tries to aggressively remove nutrients from the mother, while the maternal genome suppresses this to protect the mother.

Suggested Reading

Paldi, A. 2003. Genomic imprinting: could the chromatin structure be the driving force? Curr. Top. Dev. Biol. *53*:115-138.

Wilkins, J.F. and D. Haig. 2003. What good is genomic imprinting: the function of parent-specific gene expression. Nat. Rev. Genet. *4*:359-368.

Feinberg, A.P., H. Cui, and R. Ohlsson. 2002. DNA methylation and genomic imprinting: insights from cancer into epigenetic mechanisms. Semin. Cancer Biol. *12*:389-398.

HIGH-THROUGHPUT SNP DETECTION TECHNOLOGIES

Linked Topics:
Allele-specific oligonucleotides
Microarrays
Modifications of DNA sequencing
Real-time PCR

Amplification refractory mutation systems (ARMS) polymerase chain reaction (PCR) is carried out in two separate reactions with the same reverse primer, but with either normal or mutant allele-specific oligonucleotide (ASO) as the forward primer. The 3′ location of the mismatch in mutant ASO prevents product formation in a sample with a normal allele, while homozygous mutation fails to form a product with normal ASO. A heterozygous individual yields products in both reactions. Using all three primers in a single tube, with the mismatched primer of different length, allows a single reaction.

Oligonucleotide ligation assay (OLA) is based on absolute dependence of ligation on complementarity for adjacently placed oligonucleotides with the template DNA. PCR-amplified target DNA is hybridized with two collinear oligonucleotide probes labeled with biotin and a fluorescent reporter tag, which are subjected to ligation. Successful ligation is detected by transfer to a streptavidin-coated microtiter capture plate and fluorescence readout. This assay can detect many mutations in a single reaction, using automated DNA sequencer and fluorescent reporter molecules.

DNA microarray for single nucleotide polymorphism (SNP) detection uses sequence-specific oligonucleotide probes encompassing all possible SNPs microprinted or electronically addressed onto a semiconductor chip in a grid pattern. Hybridization of PCR-amplified fluorescent-tagged target DNA reports on the presence of or absence of particular SNPs. Real-time PCR has advanced this process significantly.

High-Throughput SNP Detection Technologies

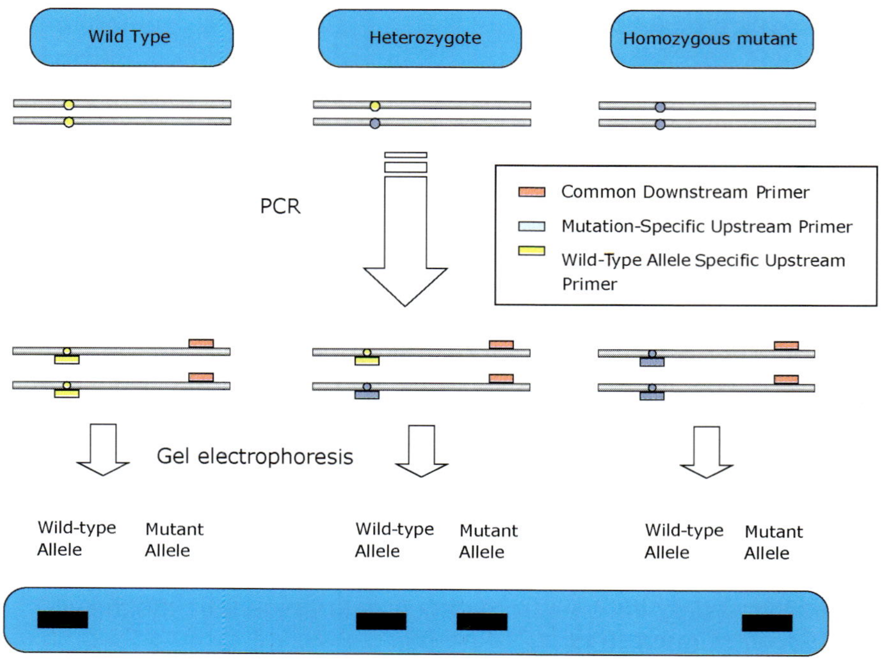

Protein Truncation Test

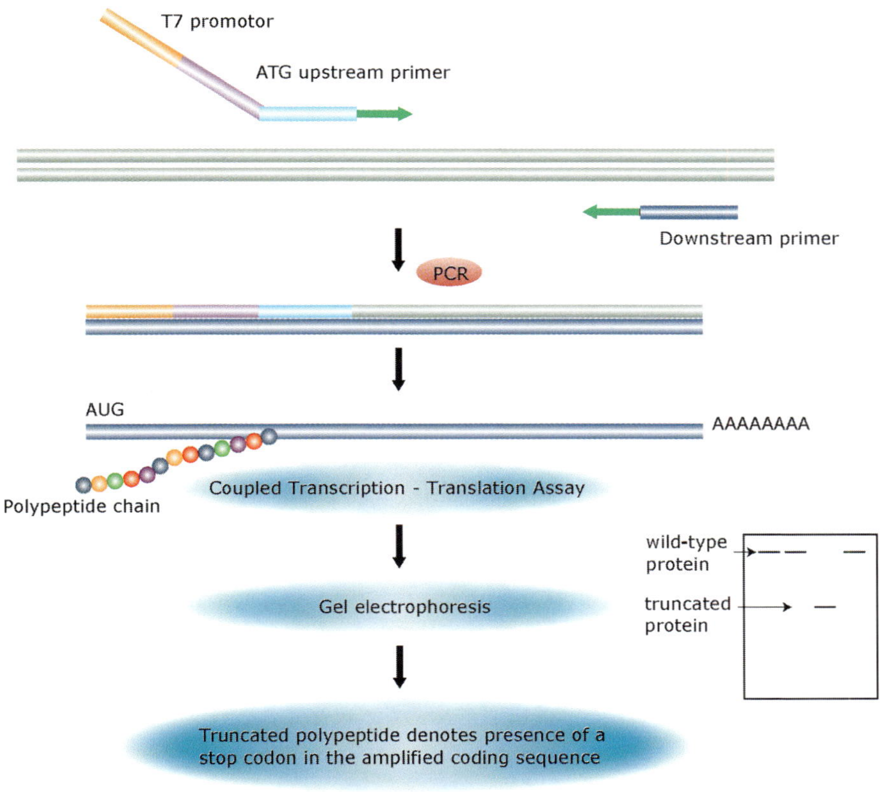

Two-dimensional electrophoresis (2-DE) allows separation of PCR products by size in one direction and by base composition in the other.

Numerous modifications of DNA sequencing technologies are being applied to SNP detection. **Mass spectroscopy (MS)** compares the ratio of molar mass to charge for target DNA sequences to identify mutations by comparing the sequence data to that of the standard fragment.

Subtractive oligonucleotide hybridization analysis is a method that uses surface plasmon resonance-based optical biosensors to detect the ASO probe remaining unhybridized in solution (by virtue of mass changes) after hybridization to immobilized target DNA.

While unable to detect missense mutations, **protein truncation tests (PTTs)** can detect nonsense, frameshift, splice site, insertion, and exon deletion mutations by virtue of translation termination of in vitro translated mRNA in a coupled transcription–translation system.

Development of competitive technologies through interdisciplinary collaborations among physical and biological researchers with emphasis on automation promises to increase the speed of SNP detection substantially in the near future.

Suggested Reading

Tsuchihashi, Z. and N.C. Dracopoli. 2002. Progress in high throughput SNP genotyping methods. Pharmacogenomics J. *2*:103-110.

HOMEOBOX GENES

Linked Topics:

DNA-binding proteins

Functional genomics

Homeobox genes are DNA-binding proteins (transcriptional factors) primarily involved in embryonic development, but also are shown to have key roles in the development of human diseases as diverse as diabetes and cancer. Several homeobox gene families are identified, all the members of which share a common sequence motif encoding 60 amino acids (**homeodomain**). (Since this conserved sequence was often depicted in boxes across homologous proteins, these genes came to be called homeobox genes.) The protein products of these genes control the patterned expression along the anteroposterior axis, directing the development of organs and structures characteristic of respective regions. Formation of specific heterodimers among the members of homeobox proteins are often required for their specificity in gene regulation. Any dysfunction of these genes due to mutations results in the development of body parts in unusual positions, often with grotesque results. Thus, for example, mutations in the gene antennapedia in *Drosophila* have been shown to cause growth of miniature legs on their heads where antennae ought to be.

The first homeiotic genes, antennapedia and ultrabithorax, were discovered in *Drosophila*, which were shown to control position-specific differentiation of body regions. Arrangement of homeobox genes into physical clusters is crucial in the patterning of body parts (phenomenon of colinearity). Peculiarly, the linear order of the genes in each cluster is the same as the order in which they are transcribed. Thus, in *Drosophila*, the gene at the 5′ end of the cluster is transcribed only in the first (anterior) segment of the fly, the next in the second segment, and so on, until the 3′ genes, which are transcribed only in the posterior segment. Evolutionarily highly conserved from organisms as primitive as hydra to the mammals, homeobox genes are so similar that human ***HOX*** genes can be transferred into *Drosophila* and can correct homeotic mutations there. There are four gene clusters in the human system (*hox a, hox b, hox c, and hox d*), each containing at least nine tightly linked homeobox genes. The recently discovered human transcription factors HOXa10 and HOXa11, and their mouse homologues, have been shown to be necessary for fetal implantation to the uterine endometrium. Implantation defects, thus, may be molecularly diagnosed by studying mutations in these genes.

Emerging technologies of functional genomics or high-throughput expression profiling are likely to yield significant information regarding molecular interactions of homeobox proteins as components of the cellular signal transduction network.

Suggested Reading

Mann, RS. 1997. Why are Hox genes clustered? Bioessays *19*:661-664.

Zhao, Y. and H. Westphal. 2002. Homeobox genes and human genetic disorders. Curr. Mol. Med. *2*:13-23.

Cillo, C., M. Cantile, A. Faiella, and E. Boncinelli. 2001. Homeobox genes in normal and malignant cells. J. Cell. Physiol. *188*:161-169.

HUMAN GENOME PROJECT

Linked Topics:
DNA sequencing
Physical mapping

The Human Genome Project is an international collaborative project with the ultimate aim of obtaining a complete physical map of the human genome by DNA sequencing. Since the human mitochondrial genome has already been completely sequenced, the genome now under investigation is the nuclear genome. The Human Genome Project was arose out of the need for a large-scale project to optimize the existing methods and develop new technologies to increase DNA mapping and sequencing efficiency by 10- to 20-fold. The project was initiated in 1990 under the auspices of the National Institutes of Health (NIH) and Department of Environment (DOE) of the United States. The objectives realized by April 2003 (2 years ahead of previous projections) are as follows:

- Construction of a high-resolution genetic map of the human genome.
- Production of a variety of physical maps of all the human chromosomes and that of the DNA of selected model organisms.
- Determination of the complete sequence of human DNA and of DNA of selected model organisms.
- Development of the capabilities for collecting, storing, and distributing the data produced.
- Creation of appropriate technologies necessary to achieve these objectives.

The fifth objective spans the first four. The Human Genome Project is now truly an international project and major programs are well entrenched in countries like the United Kingdom, France, Japan, Denmark, Germany, Italy, the Netherlands, China, Korea, Australia, and Canada. The Human Genome Organization, or **HUGO**, was established to coordinate the different international efforts, to facilitate exchange of international research resources, and to encourage public debate and advise on the implications of human genome research. The different HUGO centers are HUGO Europe, based in London, HUGO Americas, based in Bethesda, and HUGO Pacific, headquartered in Tokyo.

The large-scale sequencing was undertaken only after significant clone contexts, spanning individual chromosomes, were acquired. In the meantime, genome sequencing of some model organisms was already undertaken and in some cases completed. The model organisms selected were mouse, *Drosophila*, *Caenorhabditis elegans*, yeast, and *Escherichia coli*. Of these, yeast, *E. coli*, and *C. elegans* have already been completely sequenced. The human genome has so far been predicted to contain 30,000–35,000 genes. The Human Genome Project has thrown open debates on a variety of **ethical, legal, and social issues** (ELSI). The main concerns include fairness in the use of genetic information, privacy of the information, and the safety and efficacy of the genetic tests. The information derived from the Human Genome Project can be very useful in diagnostics, preventive medicine, gene therapy, and the understanding of the basic biological defects based on the gene information, which in turn would lead to rational drug design as envisaged in pharmacogenomics.

Suggested Reading

Hopkinson, I. and J. Mackay. 2002. The clinical impact of the Human Genome Project: inherited variants in cancer care. Ann. Oncol. *13* (Supp. l4):105.7.

Zneimer, S.M. 2002. The human genome project: exploring its progress and successes and the ethical, legal, and social implications. Clin. Leadership Manag. Rev. *16*:151-157.

IMMUNO-PCR

Linked Topics:

Modifications of PCR

PCR and PCR optimization

In conventional antigen detection, a primary antibody binds the antigen and a secondary antibody directed against the primary is attached to enzyme labels [as in immunohistochemistry or enzyme-linked immunosorbant assay (ELISA)] or radioactive labels [as in radioimmunoassay (RIA)], which serve as signal reporters. Immuno-polymerase chain reaction (PCR) is a similar antigen detection system, where PCR is used to amplify marker DNA sequences that have been attached specifically to antigen–antibody complexes via a bifunctional molecular linker. The marker DNA is amplified by PCR using appropriate primers. The detection of a specific PCR product indicates the presence of the antigen to start with.

A unique and versatile bifunctional molecular linker is a streptavidin–protein A chimera. Streptavidin is a protein produced by *Streptomyces avidini*, which specifically binds biotin with an extremely high affinity (kd=10^{-15}), whereas protein A is a cell wall constituent of *Staphylococcus aureus*, which specifically binds F_c domains of immunoglobulin (Ig)G molecules without affecting their antigen binding. This chimera, therefore, can bind to biotin on one hand and to the IgG on the other, thus allowing specific conjugation of any biotinylated DNA molecule to antigen–antibody complexes. Alternatively, using a biotinylated secondary antibodies, a biotinylated marker DNA can be attached using streptavidin or avidin linker.

Microtiter plate format is the most common manner for immuno-PCR, where the antigen is immobilized on the well surface, followed by blocking of free binding sites on well surfaces, antigen-primary antibody-secondary antibody binding, and binding of the linker chimera prebound to biotinylated marker DNA. PCR is then carried out with appropriate primers, and the products are analyzed. Both monoclonal and polyclonal antibodies can be used in immuno-PCR. Generally, very low concentrations of antibody and marker DNA are sufficient, due to the extremely high sensitivity of immuno-PCR. While a small number of samples can easily be analyzed by gel electrophoresis, direct incorporation of fluorochromes into PCR products provides for high-throughput analyses as in clinical diagnostics. Use of immuno-PCR can allow analysis of an antigen, even at single cell levels due to high sensitivity, which otherwise may not be possible owing to low-level antigen expression, as in the case of detection of rare antigens at early stages of diseases or infectious development.

The efficient and complete removal of unbound and nonspecifically bound antibody and marker DNA is the key factor in this protocol, so as to minimize background signals. Physiological fluids, such as blood samples, serum, etc., containing high levels of endogenous Igs can generate high background signals. The marker DNA molecules and their primers ought to be changed frequently to avoid generation of false positive signals due to contamination. This also offers easier control of false positive signals compared to any other PCR-based detection method.

Immuno-PCR

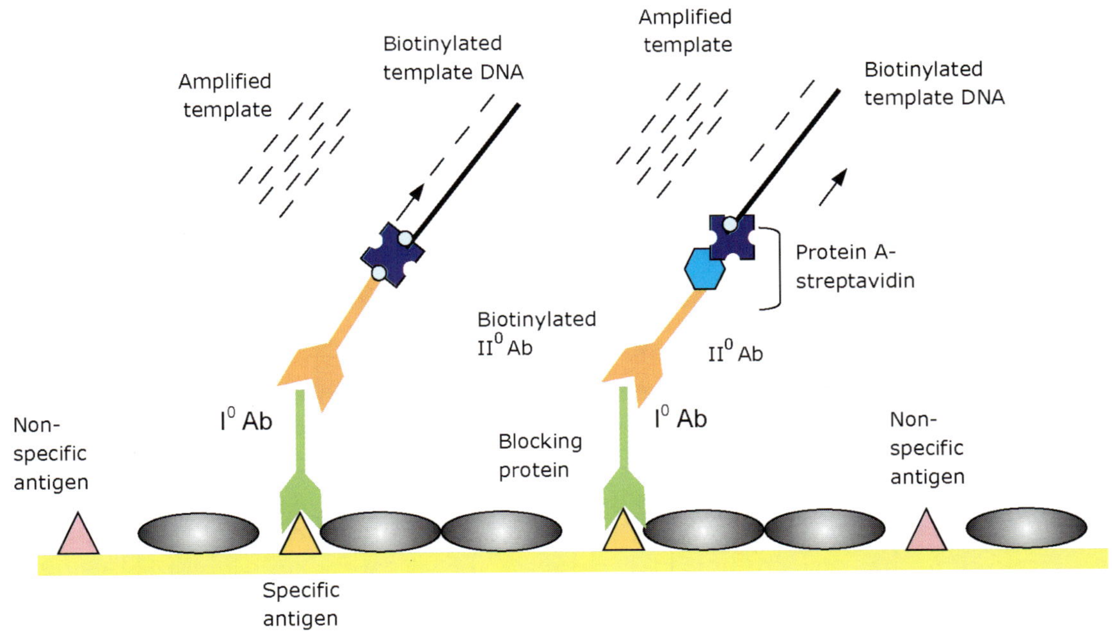

Facility for automated assay systems and control over specificity, which comes from the ability to change the DNA markers, make immuno-PCR an attractive antigen detection system, even for a single antigen molecule.

Suggested Reading

Cao, Y., K. Kopplow, and G.Y. Liu. 2000. In-situ immuno-PCR to detect antigens. Lancet *356*:1002-1003.

Case, M.C., A.D. Burt, J. Hughes, J.M. Palmer, J.D. Collier, M.F. Bassendine, S.J. Yeaman, M.A. Hughes, and G.N. Major. 1999. Enhanced ultrasensitive detection of structurally diverse antigens using a single immuno-PCR assay protocol J. Immunol. Methods *223*:93-106.

Case, M., G.N. Major, M.F. Bassendine, and A.D. Burt. 1997. The universality of immuno-PCR for ultrasensitive antigen detection. Biochem. Soc. Trans. *25*:374S.

IN SITU HYBRIDIZATION

Linked Topics:

Autoradiography

Bacterial artificial chromosomes

Bacteriophage

Human Genome Project

Modifications of fluorescence in situ hybridization

Nonisotopic labeling

Nucleic acid hybridization

P1-derived artificial chromosomes

Physical mapping

Yeast artificial chromosomes

Based on the basic principles of nucleic acid hybridization, the various forms of in situ hybridization (ISH) permit detection of target nucleic acid sequences within a particular cellular compartment. These are tissue ISH, chromosome ISH, and fluorescence ISH.

Tissue ISH: In this procedure, a labeled probe is hybridized against RNA in tissue sections. The tissue sections are made from either paraffin-embedded or frozen tissue using cryostat and then mounted onto glass slides. A hybridization mixture, including labeled nucleic acid probe, is applied to the section of the slide, covered with a glass coverslip, and allowed to hybridize in presence of typically as much as 50% formamide. Single-stranded complementary RNA (cRNA) probes (**riboprobes**) are preferred, which are generated by cloning a gene in M13 bacteriophage-based transcriptional vectors. High-energy radioisotopes such as ^{32}P are not suitable for probe labeling, because they cause signal scattering. Hence, the radioisotope of choice in earlier ISH experiments has been tritium (H^3). However, the very weak emission of H^3 requires prolonged exposure times, which again leads to increase in background noise. Thus, application of nonisotopic labeling and detection methods have replaced use of radioactive probes. Following hybridization and washes to remove excess labeled probe, the signal is visualized using autoradiography.

Chromosome ISH: This procedure uses suitably labeled DNA probe to hybridize with in situ denatured chromosomal DNA and detect the location of the complementary target DNA sequence. Briefly, an air-dried metaphase or prometaphase chromosome preparation (usually from peripheral blood lymphocytes or cultured lymphoblastoid cells) is treated with RNase and proteinase K to expose chromosomal DNA. This is then denatured by formamide treatment and hybridized in situ with a labeled nucleic acid probe. Chromosome banding, performed either before or after the hybridization, allows the correlation of the signal with the chromosome band pattern and the identification of a map location for the target. The chromosomal ISH technique can discern DNA sequences situated several megabases away from each other.

Fluorescence ISH (FISH): This molecular cytogenetic technology has proved to be a pivotal tool in physical mapping and in the progress in the Human Genome Project. FISH involves use of several kbp-long **locus-specific** or **alphoid (centromeric) repeat DNA probes**, derived from DNA cloned in cosmid, bacteriophage, bacterial artificial chromosome (BAC), P1-derived artificial chromosome (PAC), or yeast artificial chromosome (YAC) vectors. (Alphoid probes are useful for identification of specific chromosomes.) These single-stranded probes are labeled by direct incorporation of fluorescent molecules (or their posthybridization interaction with fluorescently tagged reporter molecules) and hybridized to metaphase chromosomes. The large interspersed repetitive DNA sequences present in these large probes necessitate the use of chromosome in situ suppression hybridization, wherein prior mixing of the denatured probe with a large excess of unlabeled total genomic DNA saturates the repetitive elements in the probe. This prevents erroneous results. The FISH results can be visually scored conveniently using an epifluorescence microscope. In metaphase spreads, double-spot positive signals imply diploidy. Sophisticated digital image analysis equipment is now routinely used to analyze the results of FISH. Conventional FISH procedures on metaphase chromosomes can resolve target DNA sequences 2–5 megabases apart, but resolution of less than a megabase is possible in more extended interphase chromosomes. Further resolution in FISH is obtainable by one of the many modifications of FISH.

Suggested Reading

Parker, R.M. and N.M. Barnes. 1999. mRNA: detection by in situ and northern hybridization. Methods Mol. Biol. *106*:247-283.

King, W., J. Proffitt, L. Morrison, J. Piper, D. Lane, and S. Seelig. 2000. The role of fluorescence in situ hybridization technologies in molecular diagnostics and disease management. Mol. Diagn. *5*:309-319.

Gozzetti, A. and M.M. Le Beau. 2000. Fluorescence in situ hybridization: uses and limitations. Semin. Hematol. *37*:320-333.

Nath, J. and K.L. Johnson. 2000. A review of fluorescence in situ hybridization (FISH): current status and future prospects. Biotech. Histochem. *75*:54-78.

IN VITRO PACKAGING SYSTEMS: PACKAGING EXTRACTS

Linked Topics:

Bacteriophages

Bacteriophage P1 vectors

Gene cloning

Lambda (λ) bacteriophage

Plasmids

Lambda (λ) bacteriophage cloning vectors can accept larger stretches of insert DNA than traditional plasmid vectors. But efficiency of introduction into a bacterium by transformation is limited because of the nearly 50-kb size, which makes it sensitive to shearing or breaking. It is possible to package the ligated DNA into viral capsids in vitro, and then infect the cells with the newly created infectious particles. Understanding the normal in vivo packaging will help in understanding the concept of in vitro packaging.

Normal in vivo packaging of λ bacteriophage DNA involves participation by phage major capsid protein product of gene E, minor capsid protein product of gene D, assembly proteins, and the enzyme **pacase**. The pacase cleaves the concatemeric phage DNA at neighboring *cos* sites, which results in unit lengths of phage DNA. During assembly, a prehead capsid structure is made by E, followed by insertion of a unit length of DNA into the prehead. The head maturation is then completed by insertion of minor capsid protein D. The assembly proteins ensure joining of the completed tails to the completed heads. A mutation in E gene results in nonproduction of protein E, as in the λ lysogen *Escherichia coli* strain carrying amber mutation E^M. This lysogen is capable of the synthesis of D protein and the assembly proteins, producing a bacteriophage with a tail but not a functional phage particle. Conversely, another λ lysogen, carrying amber mutation D^M, can allow synthesis of bacteriophages with λ tails and of proteins E and assembly proteins, but it is deficient in functional head formation due to lack of protein D. Mixing the lysates (extracts) of these two lysogens in the presence of recombinant DNA (insert DNA flanked by *cos* sequences) permits complementation of their individual missing functions of E and D proteins, leading to the assembly of functional head and phage particles. There are several commercial sources of "packaging extracts" that permit packaging ligated λ concatamers into viral capsids. Such packaging extracts can effectively insert up to 40 kb of insert DNA and have proved to be a very valuable tool for gene cloning.

To further overcome insert size limitation of the λ bacteriophage systems, a novel method for cloning and packaging DNA fragments using a bacteriophage P1 vector system was developed, which allowed cloning of genomic DNA fragments as large as 70–95 kb with high efficiencies. A vector such as SacBII is digested with restriction endonucleases *Sca*I and *Bam*HI, and the ends are dephosphorylated. Thus, two vector "arms" are generated: the *Sca*I-*Bam*HI short arm and the *Sca*I-*Bam*HI long arm. Genomic DNA are partially digested with *Bam*HI or another compatible end restriction endonucleases. Size-fractionated fragments between 70 and 95 kb are ligated to the vector arms, generating a series of linear molecules. Ligation between two short arms results in molecules containing neither the origins of replication nor the kan^R (gene for kanamycin resistance) gene, while ligation products of both long arms will exclude the pac site (hence no packaging into the phage heads), making both kinds of products nonviable. The only viable recombinants will be those that consist of the insert sequence flanked by a short and a long arm. Phage P1 can accommodate a total DNA length of approximately 110–115 kb. Inclusion of the distal loxP site in the recombined DNA is essential for its circularization upon injection into the host bacterial cell by its cre+ protein. Since inserts longer than 95–100 kb will truncate without the loxP site, such recombinants will fail to circularize upon injection into the host. Once the recombinant P1 phage genomes of the appropriate size are injected and circularized in the host, their DNA replicates using the plasmid origin of replication. Propagation of cells on sucrose-containing media only permits growth of colonies with genomic DNA inserts (positive selection).

Suggested Reading

Yang, Q. and C.E. Catalano. 2003. Biochemical characterization of bacteriophage lambda genome packaging in vitro. Virology *305*:276-287.

KNOCK-OUT MICE

Linked Topics:
Gene targeting
Reverse transcription PCR
Southern hybridization

These animal models are a specialized category of genetically engineered transgenic animals. Pathogenic gene mutations can either be dominant or recessive. The dominant mutations result in gain of function and can cause disorders even in the heterozygous condition, while the recessive mutations must be homozygous to result in disease. The latter are thus characterized by loss of function. The animal models for autosomal recessive diseases, therefore, ought to have both the alleles of the relevant gene "knocked out" by either alteration, which would render them inactive, or replaced by an irrelevant gene. Since mice are, in many cases, the mammalian models most easily amenable to genetic manipulation that faithfully reflect human pathophysiology, the knock-out mice are becoming an excellent system to understand human gene functions.

A selectable replacement gene (such as the neomycin resistance gene) is substituted for the target gene by gene targeting via targeting plasmid vectors. The methodology of site-specific homologous recombination is employed to install the replacement gene in the **embryonic stem (ES) cells**, which are then microinjected into the mouse blastocyst. If the ES cells contribute to germ cells of the embryo, they transmit the entire haploid genome, including the replacement gene in place of the knocked-out gene, to the subsequent generations. The heterozygotes are then bred with each other to yield homozygotes lacking both the alleles of the target gene. In addition to the positively selectable replacement gene, anoth-

Knock-Out Mice

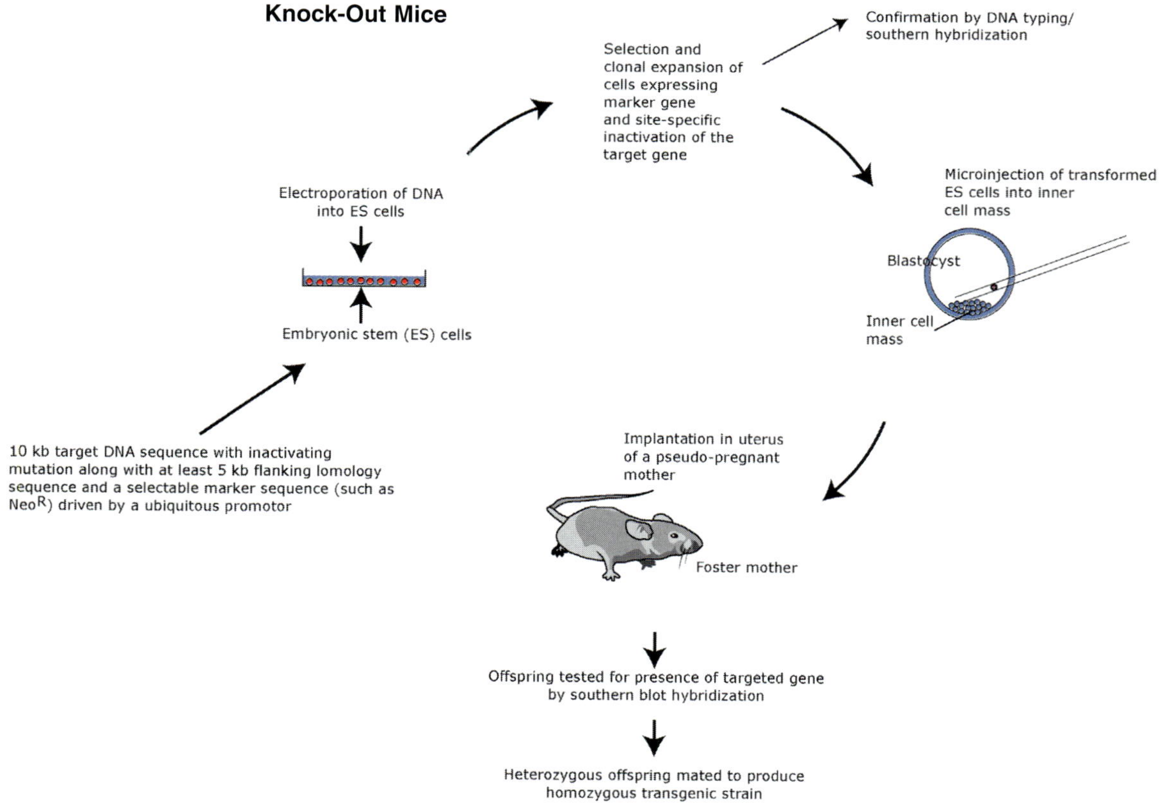

er selectable gene like viral thymidine kinase is included downstream of the flanking homologous sequence, which permits negative selection of any nonsequence-specific uptake of genetically engineered DNA. The success of contribution of ES DNA to the knock-out mouse can be ascertained visually by choosing the ES cells and blastocysts from two different strains of mice whose genes for coat color differ. Functional inactivation of the targeted gene is demonstrated by Southern hybridization, reverse transcription polymerase chain reaction (RT-PCR), and protein assays to detect and quantify the activity of the transcript and protein product of the targeted gene. Loss of function of some targeted genes (like mediators of inflammation, for example) may not become apparent until a specific initiation event (such as inflammation) occurs first. For some inactivated genes on the other hand, quite unexpected phenotype might lead to the identification of a new role for the gene product.

The involvement of a given gene in important processes such as fetal development and organogenesis can be studied by using knock-out mice deficient in that gene. Genetic deficiencies can be intermingled by cross-breeding mice that are knocked-out for different genes, permitting the study of overlapping gene functions. The technology of knockout-mice is also useful to understand the underlying mechanism of diseases like Huntington's disease, where, although the knock-out resulted in fetal death of the mice and not the expected postnatal neural degeneration, the role of extent of repeat expansion in disease onset was understood. This technology also provides useful disease models to assess the efficacy of new therapeutic chemicals.

The limitations of this technology include possible prenatal or neonatal deaths in mice as a result of loss of the gene function, which precludes study of that gene further in development. A polygenic phenotype or disorder is also unlikely to be apparent if only one gene in the pathway is knocked-out. One significant limitation is that, despite extensive genetic similarities between mice and humans, some murine proteins might not fulfill the function of the target protein in humans, possibly via subtle but crucial structural differences. In such cases, gene knock-out technology may not be directly applicable to human disease. Alternatively, the biological redundancy due to the existence of multigene families can result in compensation of the targeted gene function by sibling proteins. It might also be difficult to generate a gene knock-out line if the targeted gene is important for fertility or the processes of reproduction, implantation, and gestation; all of which result in a high fetal loss despite the initial success in breeding homozygous knock-out animals. Nevertheless, a better understanding of genetics and the immunology of mice and improvements in knock-out technology, allowing temporally controlled and tissue-specific activation of targeting vectors, could make the technology immensely useful in fulfilling its promise of associating genetics with pathogenesis.

Suggested Reading

Sora, I., K. Ikeda, and Y. Mishina. 2003. Receptor knock-out and gene targeting. Generation of knock-out mice. Methods Mol. Med. *84*:205-216.

MICROARRAYS

Linked Topics:
Autoradiography
DNA sequencing
Expressed sequence tags
Functional genomics
High-throughput SNP detection technologies
Oligonucleotide synthesis
Pharmacogenetics and pharmacogenomics

Today's DNA array technology reverses the approach of screening an array of unknowns with a defined probe [such as a cloned gene, polymerase chain reaction (PCR) product, or synthetic oligonucleotide] by positioning defined "probes" into cells of the array, and the array is then probed with the unknown sample. An array may contain all possible combinations of oligonucleotides (8- to 50-mers) or cloned cDNAs. Microarrays are a combination of technologies for creating and localizing probes, interrogation chemistry, scanning, analysis, and data management. DNA arrays are mostly employed in two applications: **genotyping** and **gene expression analysis**. Genotyping arrays are being used in high-throughput single nucleotide polymorphism (SNP) detection technologies.

Commercial **spotters** and **gridders** are available for creating DNA arrays. The robotic spatial resolution is currently the limiting factor for the density of the array. Contact microprinting, in situ oligonucleotide synthesis, or electrochemically guided probe addressing are the main options for laying down probe molecules. The surface of a glass wafer, a nylon membrane, or a silicon chip is used as a substrate.

Expression arrays are based on use of longer fragments of synthetic or complementary DNA as probes. Some arrays are designed to detect expression of a specific gene or gene families, while others can be used to query entire genomes of smaller organisms. Some arrays are even available for as many as 25,000 human cDNAs, which correspond to known genes as well as uncharacterized expressed sequence tags (ESTs).

Hybridization of the test RNA samples (which are radiolabeled or fluoroscently labeled) to the cDNA stationed in the array is followed by washing away of unbound target RNA, and the chip is ready to be scanned. Detection is effected by autoradiography or fluorometry. Bioelectronic detection, where positive events are detected by electron transfer reactions from the DNA to the substrate, or matrix-assisted laser desorption ionization time-of-flight (MALDI-TOF) mass spectrometry to analyze DNA fragments captured in an array, are some advanced options.

Most image capturing instruments use a scanning detector like the online scanning detector systems for DNA sequencing instruments, in which a multiple-point detector can sample across the array and the row-to-row step intervals. Since the microarray probe cells have continued to get smaller (currently <25 μm), detector resolution must enhance simultaneously. Signal-to-noise ratio of the hybridization can be affected by the background signal due to the fluorescence from the support matrix. Data analysis and management is a daunting task, and several companies are competing to develop appropriate software products for DNA arrays.

Soon, as a part of the worldwide functional genomics effort, the expression levels of thousands of genes will be documented for numerous developmental and disease states, further giving momentum to the fields of pharmacogenetics and pharmacogenomics.

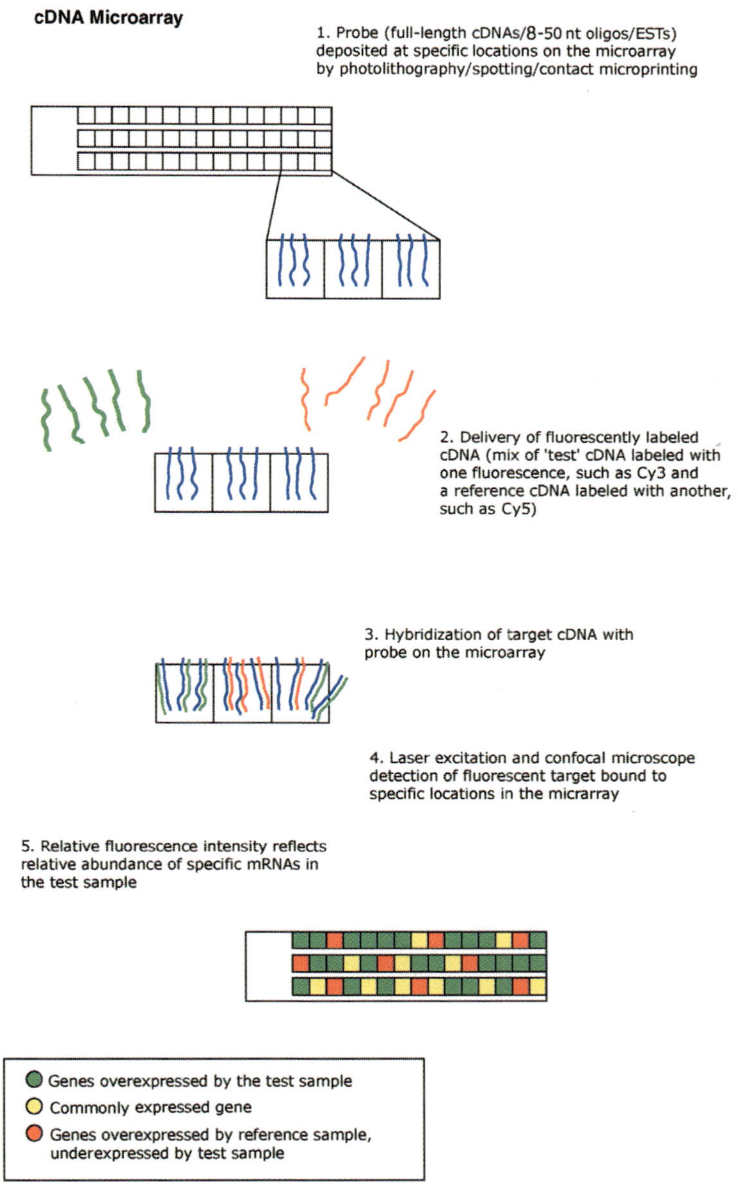

Suggested Reading

Maughan, N.J., F.A. Lewis, and V. Smith. 2001. An introduction to arrays. J. Pathol. *195*:3-6.

Nguyen, D.V., A.B. Arpat, N. Wang, and R.J. Carroll. 2002. DNA microarray experiments: biological and technological aspects. Biometrics *58*:701-717.

Hatfield, O.W., S.P. Hung, and P. Baldi. 2003. Differential analysis of DNA microarray gene expression data. Mol. Microbiol. *47*:871-877.

Conway, T. and O.K. Schoolnik. 2003. Microarray expression profiling: capturing a genome-wide portrait of the transcriptome. Mol. Microbiol. *47*:879-889.

Valafar, F. 2002. Pattern recognition techniques in microarray data analysis: a survey. Ann. NY Acad. Sci. *980*:41-64.

MODIFICATIONS OF DNA SEQUENCING

Linked Topics:

Chromosome painting

Microarrays

Nucleic acid hybridization

Oligonucleotide synthesis

Sanger's dideoxy chain termination method

Single nucleotide polymorphism detection

Cycle sequencing: This is currently a favored mode of custom sequencing, wherein very small amounts of template DNA can be sequenced by a polymerase chain reaction (PCR)-based modification of Sanger's dideoxy chain termination method. The target DNA fragment is inserted and cloned in a suitable vector, and following denaturation and sequencing primer annealing, DNA amplification is carried out using thermostable DNA polymerase and an appropriate mixture of unmodified deoxynucleotide triphosphates (dNTPs) and dideoxynucleotide triphosphates (ddNTPs). Typically, the researcher submits template DNA and primers to a core sequencing service or a custom sequencing facility, and after dideoxy sequencing in the one-sided PCRs, the data file with sequence (both as fluorogram and text strings) is sent back to the researcher by e-mail attachment.

Primer-walk sequencing: Akin to chromosome walking, this method employs subsequent sequencing primers derived from the prior reads until the complete insert is sequenced for long cloned fragments.

3′ exonuclease sequencing: Using one of the target DNA strands as a template, a primer extends a copy using the four nucleotides, each tagged with a different fluorochrome. The extension product is then cleaved using 3′ exonuclease sequentially, whereby the order of released fluorochromes signals the nucleotide sequence.

Sequencing by hybridization (SBH): An outgrowth from the developments in microarrays, the chip-based SBH system has all the possible combinations of sequence in a given length prepared by oligonucleotide synthesis immobilized on a chip. The target DNA fragment to be sequenced is restriction endonuclease-digested, fluorescently labeled, and hybridized with the oligonucleotides immobilized on the chip. The pattern of fluorescence bound to the nested sequence reflects the sequence of the target DNA. Remarkably, this system permits analysis of any DNA sequence in only one chip. However, the limitations include ambiguity due to target lengths and the inability to correctly distinguish direct and inverted repeat sequences owing to the inherent shortcoming of nucleic acid hybridization. The longer the target DNA, the longer are the oligonucleotides that are needed to eliminate ambiguities, which is severely limited by the number of probe cells needed to accommodate all possible combinations of nucleotides in an oligonucleotide.

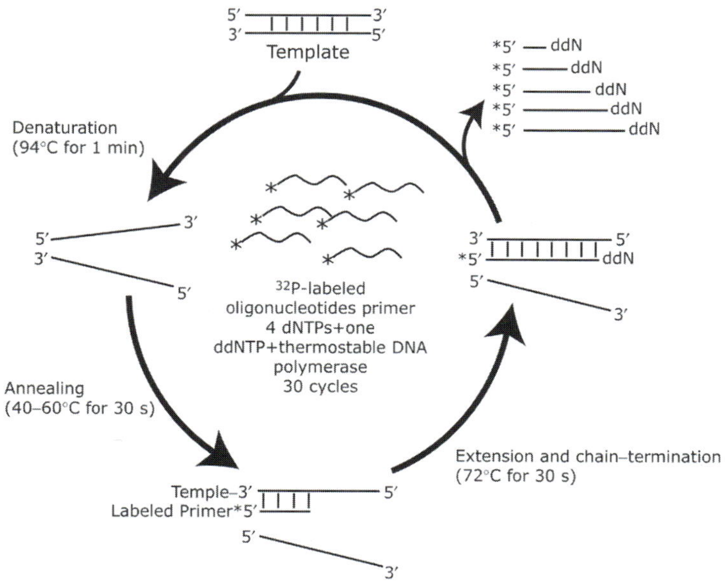

Cleavase fragment length polymorphism (CFLP): CFLP technology was developed by Third Wave Technologies, Inc. (Madison, WI, USA) and is based on the observation that single strands of denatured DNA fold into unique conformations based on their primary sequence, and certain enzymes (termed cleavases) recognize and cleave the junctions in unique locations in these secondary structures. The resulting fragments are separated in electrically charged gels, which are smaller than the conventional electrophoretic gels, thus making the process faster. The resulting folding and cleavage pattern can be read directly from the gel, yielding a signature like a bar code for every unique DNA sequence. It is noteworthy that this method reports on the target DNA sequence, not from the primary but from the secondary structure, and can also be applied for single nucleotide polymorphism (SNP) detection.

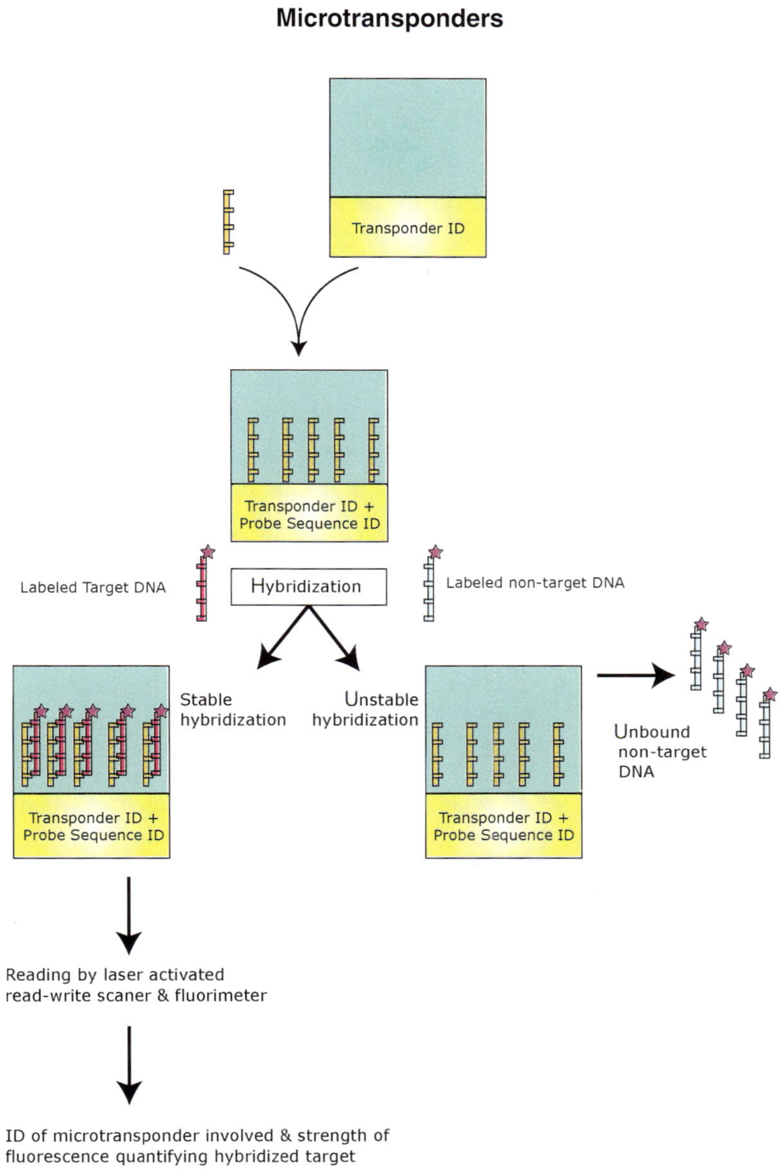

Microtransponder-based DNA sequencing: This novel method (developed by Pharmaseq, Monmouth Junction, NJ, USA) for determining DNA (or RNA) sequence involves treating the test sample with solid particles with an oligonucleotide probe attached to the surface and a transponder within. The transponder associated with the particle includes radio frequency memory elements, which indicate the sequence of the probe on its surface. Stable hybridization of denatured fluorescently labeled target nucleic acid molecules to the probes on the particle surface is followed by high-throughput screening of the particles by a fluorimeter and a dedicated read–write scanner. The fluorimeter reports on the strength of signal positivity, while the scanner activates the transponder's memory by laser to rapidly identify which microtransponder(s) are involved in the reaction, thereby identifying the sequence of the target.

Suggested Reading

Surdhar, G.K. 2002. Cycle sequencing of PCR products. Methods Mol. Biol. *187*:65-72.

Blakesley, RW. 2001. Cycle sequencing. Methods Mol. Biol. *167*:101-112.

Linked Topics:
Amplification refractory mutation sensitive PCR
Differential display reverse transcription PCR
Gene cloning methodologies
Hot-start PCR
Immuno-PCR
Nested PCR
Reverse transcription PCR
Sequencing
Southern hybridization
Touch-down PCR

Arbitrarily primed polymerase chain reaction (**AP-PCR**), also known as randomly amplified polymorphic DNA PCR (**RAPD-PCR**), is a method for creating genomic fingerprints from samples for which there is insufficient target sequence information. Short, typically 10- to 20-nucleotide-long arbitrary oligonucleotide sequence primers are employed to perform the first few PCR cycles at low stringency, generating strain-specific arrays of DNA fragments (fingerprints). Subsequent PCR cycles are then completed at a higher stringency. DNA amplified in this way is used to determine the homology or variability between species.

Allele-specific PCR (**AS-PCR**): This modification utilizes pairs of primers, one of which is an allele-specific oligonucleotide (ASO) as in amplification refractory mutation system PCR (ARMS-PCR), while the other is a sequence common to both the alleles. The amplification can thus identify and distinguish the different alleles in the given DNA sample.

Asymmetric PCR is an adaptation for generating predominately single-stranded DNA using PCR amplification, where the reaction is performed with one of the amplification primers present in nearly 100-fold molar excess.

Inverse PCR (**IPCR**) is a clever modification of PCR that allows isolation and cloning of novel hitherto unknown DNA sequence adjacent to a known sequence in genomic DNA. Here, the primers are designed to map in a tail-to-tail fashion in the known DNA sequence, about 200 nucleotides inside the end. The genomic DNA is digested using a 6-bp cutter restriction endonuclease, which does not cleave between the primers or between the distal primer and the end of the known sequence. The digested DNA is circularized using DNA ligase, and PCR is carried out. The unknown sequence thus captured can then be cloned and sequenced.

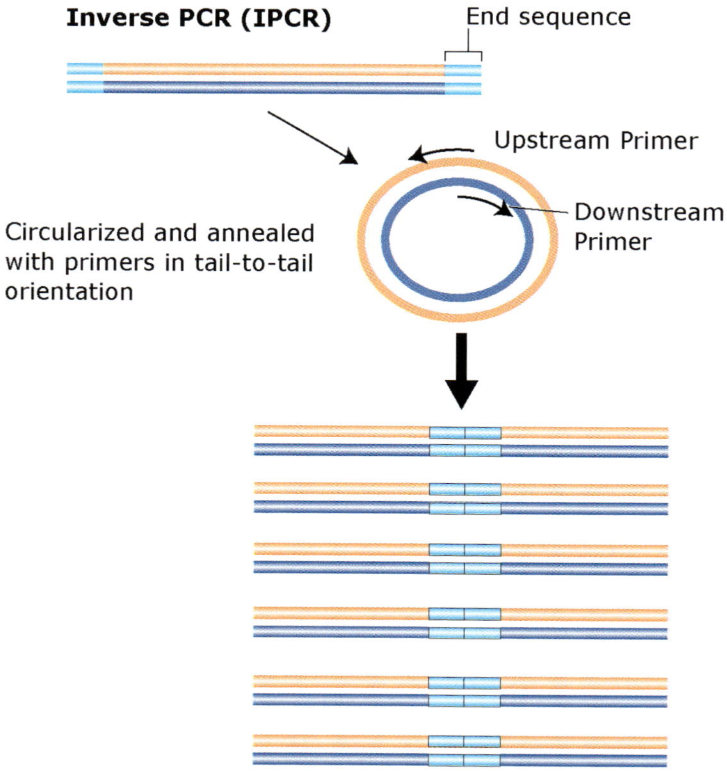

Ligase chain reaction (**LCR**): In this variation of PCR, consecutive short oligonucleotide sequences that end or begin at polymorphic nucleotides are chosen as a joint primer pair to anneal with the template DNA. The addition of thermostable DNA ligase results in the formation of a covalent phosphodiester bond between the two oligonucleotides only if the adjacent 3′ and 5′ nucleotides are perfectly complementary. Further exponential PCR amplification ensues using two additional primers complementary to the product of the first ligation reaction. Any mismatches at the termini to be ligated provide inefficient substrates for the ligase, thus resulting in failure to amplify.

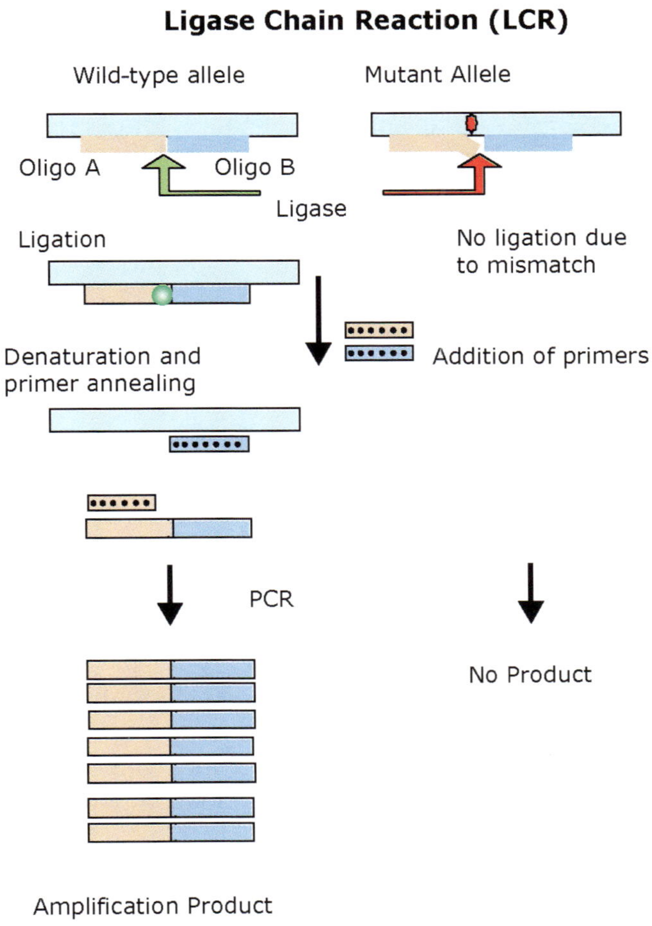

Ligase-mediated PCR (**LM-PCR**) or ligation-anchored PCR (**LA-PCR**) is often used for cloning, footprinting, or sequencing of DNA. Following cleavage of chemically modified genomic DNA, a specific anchor sequence is annealed to the cleaved fragments and extended by a DNA polymerase. A synthetic linker is ligated to the resulting blunt-ended molecules. Subsequent PCR products are generated using the linker–anchor primers and gene-specific primers and are detected with a gene-specific probe in Southern hybridization.

Interspersed repetitive sequence PCR (**IRS-PCR**): Here, primers specific for a human's long interspersed repeat elements are used to amplify and analyze genome stretches and can be combined with Alu-specific primers (**Alu-PCR**) to enhance the applicability.

Other modifications include reverse transcription PCR (RT-PCR), differential display reverse transcription PCR (DDRT-PCR), amplification refractory mutation sensitive PCR (ARMS-PCR), immuno-PCR, touch-down PCR, gradient PCR, hot-start PCR, and nested PCR.

Suggested Reading

Hui, E.K., P.C. Wang, and S.J. Lo. PCR-based strategies to clone unknown DNA regions from known foreign integrants. An overview. Methods Mol. Biol. *192*:249-274.

MODIFIED OLIGONUCLEOTIDES

Linked Topics:

Antisense oligonucleotides

The antisense oligonucleotides (oligos) synthesized using natural nucleosides are highly hydrophillic and anionic in nature, yet are taken up rapidly by the cells. This uptake is probably receptor-mediated. But effective concentration of these oligos in cytoplasm is limited by their being confined to lysosomes or endosomes. In addition, these oligos are degraded in serum within few hours due to 5′- and 3′-exonucleolytic activity. A number of structural modifications have, therefore, been made to the natural β anomers to improve their stability and effective intracellular concentration. These modifications are created in either the phosphate backbone, bases, or sugar moieties. Dictated by the ultimate function desired, either transcription inhibition or translation inhibition, an oligo can be designed as triplex-forming or antisense-inhibitory, respectively.

Backbone modifications: To increase the stability of the oligos while retaining their specificity, various changes in backbone have been attempted. **Phosphorothioates** (sulfur in phosphate moiety) and **methylphosphonates** (a methyl group in phosphate moiety) are the most common modifications to the phosphate backbone. **Chimeric** antisense molecules that combine methylphosphonate residues with either phosphodiesters or phosphorothioates have also been in use and show significant promise. The phosphate-sugar backbone can be completely replaced by incorporating components such as peptide chain [to form peptide nucleic acids (**PNAs**)] or carbamate-linked morpholino chain (morpholinonucleosides). The PNAs have high nuclease resistance and superior target binding affinity, but are deficient in penetration into the cells.

Sugar and base modifications: The natural deoxyribose sugar can be replaced by either α-deoxyribose, β-L-deoxyribose, or 2′-modified ribose moieties. This increases the binding affinity of the oligos to RNA, while resisting endonucleolytic cleavage. The preferred substitutions in place of natural bases are 5-methyl cytosine or 2,6-diamino-purine. These changes can make an oligo more effective vis-a-vis triple helix formation function.

Hybrid oligos have also been synthesized containing stretches of deoxyribonucleotides alternating with ribonucleotides where 2′-O-methyl amidites are used for RNA-like portions. Besides these alterations, the 5′ and 3′ termini of the oligos can be modified by conjugation with agents like cholesterol or poly-L-lysine (to increase lipophilicity and, thereby, increase transmembrane transport) and psoralen or acridine (to improve target binding affinity).

Toxicity: Since antisense oligos are more toxic in rodents than primates, nonhuman primates are used in toxicity studies prior to human preclinical trials. Although preliminary data from these and ongoing human clinical trials indicates that none of the advanced modifications to the oligos are acutely toxic, phosphorothioate oligos (the most widely used antisense agents) have been shown to interact in a sequence-independent manner with many functional proteins such as p53, protein kinase C, H-Ras, and

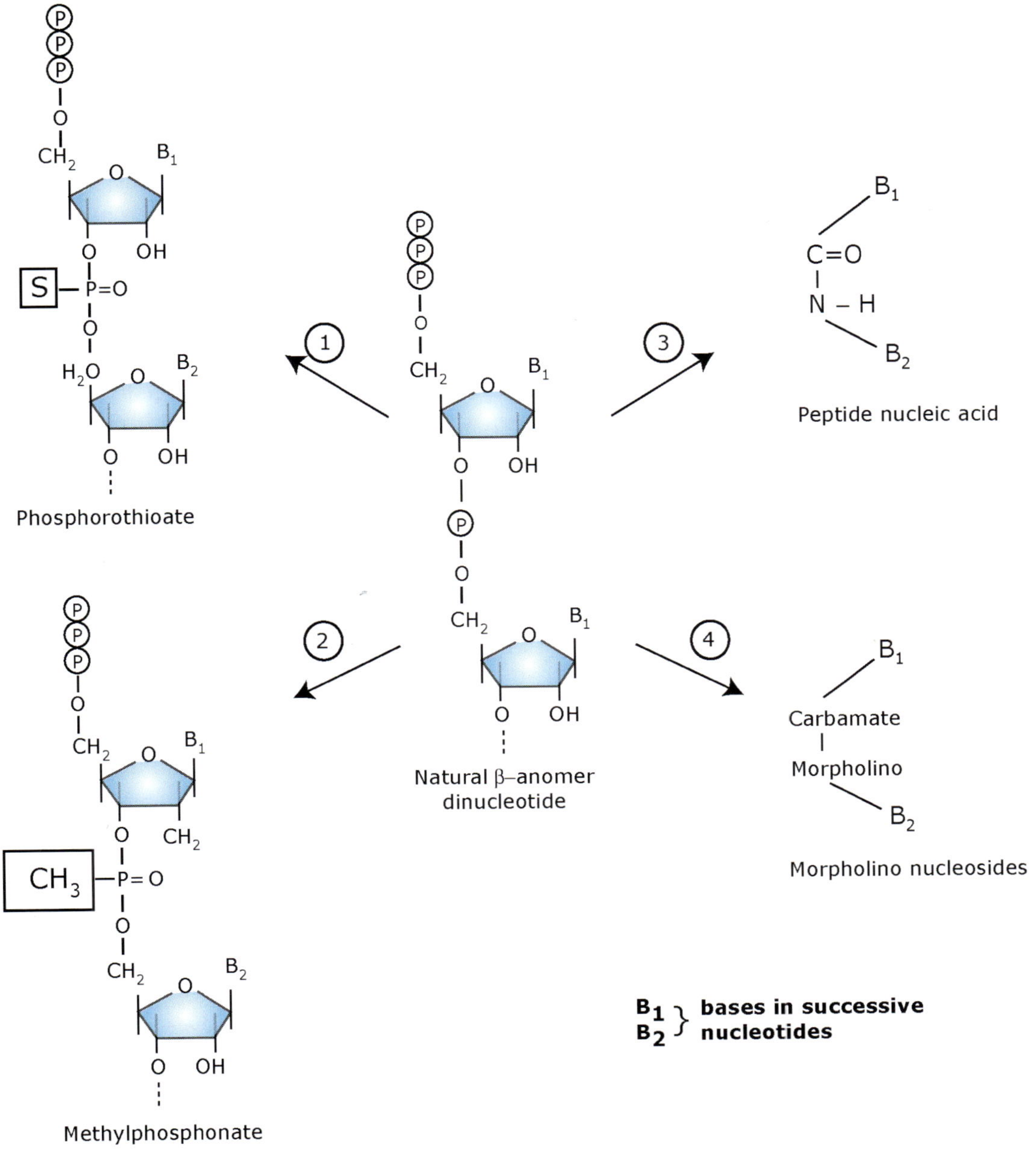

thrombin. They also seem to activate complement pathways and could be potentially immunogenic due to changes in backbone structure, charge, and molecular size—perhaps the reason for increased immunoglobulin (Ig)G and IgM production, lymphoid infiltration, and necrosis in kidneys, liver, and heart. Efforts are on to decrease these sequence-independent side effects, while increasing the specific tissue delivery and efficacy.

Suggested Reading

Micklefield, J. 2001. Backbone modification of nucleic acids: synthesis, structure and therapeutic applications. Curr. Med. Chem. *8*:1157-1179.

Manoharan, M. 1999. 2'-carbohydrate modifications in antisense oligonucleotide therapy: importance of conformation, configuration and conjugation. Biochim. Biophys. Acta *1489*:117-130.

MOLECULAR CHAPERONES

Linked Topics:
DNA-binding proteins

Molecular chaperones are the proteins that participate in important cellular functions of protein folding and intracellular protein trafficking. As the word chaperone suggests, these proteins act as custodians, escorts, and guardians for the polypeptide chains from the time they are synthesized until they are finally used at their destination within the cell. Thus, the chaperones prevent illicit interactions between the polypeptide chains and other biomolecules and help guide these polypeptide chains to their end-use location.

The chaperones were initially identified as proteins induced following heat shock, hence called **heat shock proteins** (hsp). Now it is appreciated that most of these proteins are synthesized even at physiological conditions and are critical for normal cell functioning. The two best-studied hsp families are hsp70 and hsp60. While the hsp70 proteins act by preventing the incorrect folding of polypeptide chains, the hsp60s, also known as chaperonins, facilitate the process of correct folding of polypeptides.

The **hsp70s** are seen across all organisms, with multigene families encoding these proteins in the eukaryotes. As a general rule, the amino terminal sequence of these proteins harbors an ATP-binding site, while their carboxyl terminus holds a peptide-binding site, recognizing a 7-amino acid sequence in the target polypeptide chains. The ADP-bound form of hsp70 enhances peptide binding, while the ATP-bound form favors peptide release from the carboxyl end. Thus, alternating cycles of binding and hydrolysis of ATP can determine the sequestration and release of polypeptides. The sequestration by hsp70s possibly prevents incorrect folding of proteins and may also facilitate translocation and proper spatial positioning of the proteins from the cytosol into the endoplasmic reticulum or mitochondria.

The **hsp60s (chaperonins)** form large ribosome-sized protein complexes comprising 14 proteins, each approximately 60 kDa, arranged into two-tier seven-member ringed structures. In association with similar two-ringed structures of 10-kDa proteins (**chaperonin-10**), the hsp60s enclose polypeptide chains that are already folded in their secondary structures and allow their proper tertiary folding in an ATP-dependent manner. The hsp60s, thus, probably aid proper folding of those proteins that are unable to do so by themselves. These may even be required for accurate assembly of multimeric enzyme complexes within the cell.

The hsp60s probably act in tandem with hsp70s, such that the unfolded polypeptide chains delivered through the membranes by hsp70s are taken up by the hsp60s for proper folding and eventual topologically correct release.

Hsp90s, yet another class of chaperones studied most extensively in mammalian systems, are bound to the inactive steroid hormone receptors. The latter are the DNA-binding proteins that, on interaction with steroid hormones, up-regulate the steroid-responsive genes. The hsp90s maintain the receptor molecules in an inactive state by masking their DNA-binding site. The release of these receptors thus allows their interaction with DNA. The hsp90s have also been shown to bind the src protein kinase.

Suggested Reading

Walter, S. and J. Buchner. 2002. Molecular chaperones—cellular machines for protein folding. Angew. Chem. Int. Ed. Engl. *41*:1098-1113.

Hartl, F.U. and M. Hayer-Hartl. 2002. Molecular chaperones in the cytosol: from nascent chain to folded protein. Science *295*:1852-1858.

Linked Topics:
Microarrays
PCR and PCR optimization
Quantitative PCR
Real-time PCR
Reverse transcription PCR

All living organisms have thousands of unique genes encoded in their genome, of which only a small fraction, perhaps 15%, is expressed in any individual cell. Therefore, it is the temporal and spatial regulations in gene expression that determine progress or impediment of life processes. Normal cellular developmental processes, as well as pathological changes that arise in diseases such as cancer, are now known to be driven by changes in gene expression. The method of mRNA differential display allows rapid, accurate, and sensitive detection of altered gene expression. Earlier alternatives to achieve this were subtractive hybridization or differential hybridization, which, though useful, were technically challenging. mRNA differential display is a form of reverse transcription polymerase chain reaction (RT-PCR), wherein the enzyme reverse transcriptase catalyzes cDNA synthesis by using a modified oligo(dT) primer, which binds the poly(A) tail of a subset of mRNAs determined by the nucleotides and incorporated at the 3′ end of the oligo(dT) primer. For example, if the oligonucleotide $(T)_{10}CA$ is used as a primer, it will preferentially prime cDNA synthesis from the poly(A) tails of those mRNAs that have a 3′ sequence ending in TG. The second (upstream) primer used is usually an arbitrary short sequence of about 10–13 nucleotides. Under conditions deliberately chosen to allow mismatches, this arbitrary primer can bind to many more sites in the RNA pools than expected for a decamer or a 13-mer. The resulting amplification patterns following PCR produce a complex ladder of bands when size fractionated in the long polyacrylamide gel. This allows side-by-side comparison of most of the mRNA classes between or among related cells. While ^{33}P-radioisotopic labels were used in earlier versions, recent application of fluorescent dyes (rhodamine derivatives) has allowed nonisotopic labeling and detection and has improved the ease and automation of the system.

The differential display method is unique in its ability to allow the visualization of all the expressed genes in a eukaryotic cell in a systematic sequence-dependent manner by using multiple primer combinations. The most interesting application of mRNA differential display is the comparison of gene expression in cells at different physiological or developmental stages. This allows identification of small subsets of genes for which expression patterns are different between the cell types. By isolating specific bands from the gel and submitting the DNA to further PCR cycles, gene clones can be obtained. Because of its simplicity, sensitivity, and reproducibility, the method of mRNA differential display, also called differential display RT-PCR or DDRT-PCR, has been utilized increasingly in wide ranging applications in developmental biology, cancer research, neuroscience, pathology, endocrinology, plant physiology, evolutionary biology, and many other fields.

While this technology is largely being replaced by quantitative PCR technologies such as real-time PCR or by cDNA microarrays, mainly on account of the ease associated with the latter, DDRT-PCR technology still compares favorably in terms of providing an open platform for new gene discovery, amount of total RNA required, coverage of genes, and cost. The limitations of the technology include substantial labor and lack of immediate knowledge of gene sequence.

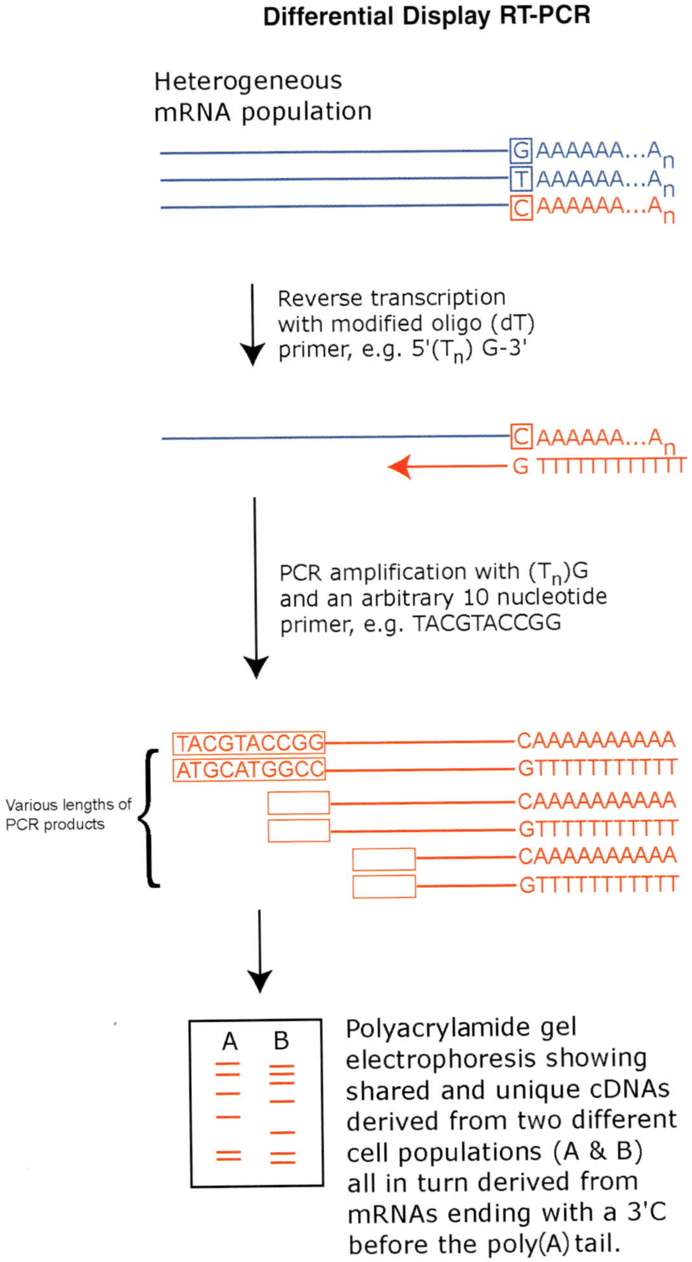

Suggested Reading

Aittokallio, T., P. Ojala, T.J. Nevalainen, and O. Nevalainen. 2000. Analysis of similarity of electrophoretic patterns in mRNA differential display. Electrophoresis *21*:2947-2956.

MULTISTEP CARCINOGENESIS

Linked Topics:
Apoptosis
Mismatch repair genes
Telomeres and telomerase

Over 90 years ago when the existence of specific oncogenes or suppressor genes was yet to be discovered, Theodore Bovari uncannily put forth the principles of oncogenesis as follows:

> In every normal cell, there is a specific arrangement for inhibiting, which allows the process of division only when the inhibition has been overcome by a spatial stimulus. Cells of tumors with unlimited growth would arise if those inhibiting chromosomes were eliminated. On the other hand, the assumption of the existence of chromosomes which promote division might also satisfy this postulate. On this assumption, cell division would take place when the action of these chromatin parts, which are as a rule too weak, should be strengthened by a stimulus, and the unlimited tendency for rapid proliferation in malignant tumor cells would be deduced.

Merely by replacing the word "gene" for the word "chromosome" in the above hypothesis, one arrives at an exact concept of carcinogenesis via activated oncogenes and inactivated tumor suppressor genes.

In the recent years, anomalies in another class of genes, the mutator genes or mismatch repair genes, have been added to the genetic factors crucial in the process of carcinogenesis. The process of carcinogenesis is often likened to a vehicle gone astray, where oncogenes are regarded as the accelerator and suppressor genes as the brakes for cell division. Any mutations in the mutator genes would result in a process of sabotage of the cellular growth control equivalent to loosening the nuts and bolts, lying in wait for disaster to strike. Cancer results from the destruction of the otherwise well-regulated genetic program, leading to uncontrolled cellular growth and proliferation. All cancers are more or less accepted to be clonal in origin. Cellular immortalization via activation of telomerase and loss of function of genes that control apoptosis (programmed cell death) may also be important events in the process of carcinogenesis.

In 1990, Fearon and Vogelstein put forth a multistep tumorigenesis model to explain the development of colorectal cancer. This model has come to be regarded as a paradigm for development of most other cancers. In colorectal cancer, malignant carcinoma develops from benign epithelial growths called adenomas. The adenomas can be classified into early adenoma (<1 cm in size), intermediate adenomas (<1 cm, but without foci of carcinoma), and late carcinomas (<1 cm, with foci of carcinoma). At the genetic level, there is no single specific sequence of mutations in the development of carcinoma from adenoma, but the most likely sequence is the one in which each consecutive step confers growth advantage on the cell. The earliest event probably is the constitutional loss of one copy of the *APC* gene at chromosome 5q21. This is followed by DNA hypomethylation, resulting in hyperproliferative epithelium. The next step, early adenoma, is characterized by the activation of *k-ras* oncogene, located at chromosome 12p. Intermediate adenoma is characterized by loss or mutation in the *DCC* tumor suppressor gene located on chromosome 18q. The loss or mutation in the *p53* gene marks the late adenoma state. These then develop into carcinoma. The mutator genes such as *hMSH2* and *hMLH1* do not play a direct role in the oncogenic pathway, but probably raise the general mutation rate during the overall transition from normal epithelium to late adenoma, thereby increasing the likelihood of genomic instability at each step along the pathway.

Although such ordered linear progression of carcinogenesis may not be a reality for all cancers (as suggested by findings in animal models in prostate cancer), the multifactorial contribution is a well-accepted phenomenon in carcinogenesis.

Suggested Reading

Minamoto, T., M. Mai, and Z. Ronai. 1999. Environmental factors as regulators and effectors of multistep carcinogenesis. Carcinogenesis *20*:519-527.

MUTATOR OR MISMATCH REPAIR GENES, MICROSATELLITE INSTABILITY, AND REPLICATION ERRORS

Linked Topics:
Apoptosis
DNA methylation
Multistep carcinogenesis
PCR and PCR optimization

General genomic instability (manifesting as chromosome rearrangements, losses, gains, and so forth) has been demonstrated in cancer cells for a long time. A peculiar kind of genomic instability was found accidentally in patients suffering with hereditary nonpolyposis colon cancer (HNPCC), which probably marks the discovery of one of the most sensational mechanisms in multistep carcinogenesis in recent years. No obvious loss of heterozygosity (LOH) was seen in chromosome regions 2p, 3p, and 7p; the loci is genetically linked with the disease. But short tandem repeat (STR) primers that amplified dinucleotide and trinucleotide repeat sequences (microsatellites) from different chromosomes by polymerase chain reaction (PCR) showed expansion or contraction of microsatellite alleles in all locations in genomes of tumor cells compared to the nontumor cells. This microsatellite instability (MSI or MIN) was shown to be due to the deficiency of DNA replication error (**RER**) correction mechanism. The genes responsible for replication error correction were located at chromosomal loci 2p, 3p, and 7p and were termed mismatch repair (**MMR**) or **mutator** genes. Such mutator genes have been known in *Escherichia coli* and yeast; in fact, the MMR genes discovered in HNPCC patients were human homologues of the corresponding yeast genes.

The microsatellite repeat sequences are prone to replication errors due to strand slippage. In yeast and *E. coli*, the MMR genes encode a repair system, which detects the mismatches and excises and replaces the incorrect bases. The 4-gene system, consisting of *Mut s*, *Mut l*, *Mut h*, and *Uvr D* (helicase II) genes, distinguishes the methylated template strand and selectively corrects the mismatch in unmethylated daughter strand. Mutations in these genes lead to several hundreds- to thousand-fold increases in overall mutation rates. In normal *E. coli* cells, the protein *Mut s* detects and binds the region spanning the mismatched nucleotide in the daughter strand. The proteins *Mut l* and *Mut h* form an ATP-dependent complex with *Mut s*, bringing a GATC sequence located within 1 kbp on either side of the mismatch into a loop close to the mismatch. *Mut h* protein cleaves 5′ to the GATC sequence on the unmethylated strand, helicase II unwinds the strands, and, depending upon whether the GATC sequence is upstream or downstream of the mismatch, a 5′ or 3′ exonuclease clips away the daughter strand to a distance of 1 to 2 kbp from the mismatch. DNA polymerase III then resynthesizes a correct copy, and DNA ligase closes the phosphodiester bond.

The loss of function of these genes is caused either by mutations or silencing of gene promoters by DNA methylation and results in accumulation of uncorrected replication errors, probably making the growth suppressor genes and oncogenes vulnerable for mutations. Thus, in HNPCC, A_{10} tracts in transforming growth factor β receptor 2 (TGFβ-R2) gene and G_8 tracts in *BAX* gene have been shown to have increased susceptibility for mismatch mutations. While TGFβ-R2 protein is a crucial inhibitor of signal transduction in cell proliferation, BAX protein is essential for apoptosis. The MMR mutations may also increase the susceptibility of the proliferating cells to carcinogens. Such replication error-prone phenotype (RER+) has now been shown in a number of other cancers, too. (The term "RER" has now been accepted as more inclusive and representative than "microsatellite instability"). A total of five human mutator genes have been identified: *hMSH 2* (human *Mut s* homologue 2) located at chromosome 2p21-22; *hMLH1*, *hPMS1*, and *hPMS2* (homologs of *E.coli Mut l* gene) at chromosomal locations 3p21.3, 2q31-33, and 7p22, respectively; and *hMSH6* (2p16). Although involvement of *hPMS1*, *hPMS2*, and *hMSH6* genes in cancers has yet to be substantiated, the *hMSH2* and *hMLH1* genes account for 95% mutations in HNPCC families.

Mismatch Repair System

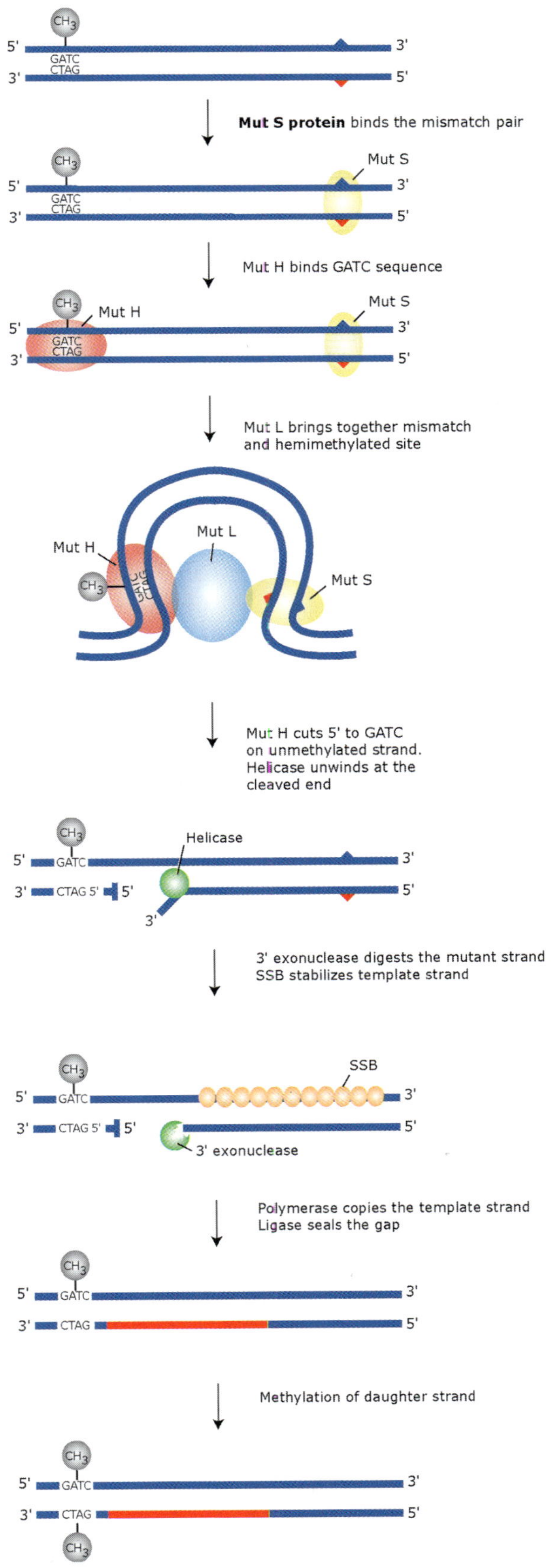

Suggested Reading

Lawes, D.A., S. SenGupta, and P.B. Boulos. 2003. The clinical importance and prognostic implications of microsatellite instability in sporadic cancer. Eur. J. Surg. Oncol. *29*:201-222.

Hussein, M.R. and G.S. Wood. 2002. Building bridges in cancer: mismatch repair and microsatellite instability. Am. J. Dermatopathol. *24*:76-81.

Wheeler, J.M., W.F. Bodmer, and N.J. Mortensen. 2000. DNA mismatch repair genes and colorectal cancer. Gut *47*:148-153.

NONISOTOPIC LABELING AND DETECTION

Linked Topics:

In situ hybridization

Quantitative PCR

Southern and Northern hybridization

Nonisotopic labeling systems involve the use of nonradioactive probes. Although only developed comparatively recently, such systems are finding increasing applications in a variety of different areas of molecular biology. Most such systems feature the chemical coupling of modified reporter molecule to a nucleotide precursor. When incorporated into DNA, the reporter groups can be specifically bound by a protein or other ligand, which has a high affinity for the reporter group. Attached to this affinity group is a marker group, which can be detected in a suitable assay.

One widely used nonisotopic labeling system utilizes the extremely high affinity of the ligands biotin and streptavidin. Biotin is a naturally occurring vitamin, whereas streptavidin is a bacterial protein that demonstrates high affinity to biotin with an affinity constant of 10^{-14}. Biotinylated probes can be made easily by including a suitable biotinylated nucleotide in the labeling reaction. A popular alternative is to use the plant steroid, digoxygenin, obtained from the digitalis plant, as a label. A specific antibody for this compound permits detection of molecules containing it. Both digoxygenin- and biotin-labeled nucleotides are designed to have an intermediate spacer so that once introduced into a labeled nucleic acid probe, the attached reporter group can protrude sufficiently far from the nucleic acid backbone so as to facilitate their detection by anti-digoxygenin antibody or streptavidin, respectively. The marker can be detected in various ways. If it carries a specific fluorescent dye, it can be detected in a fluorimetric assay. Alternatively, an enzyme such as alkaline phosphatase can be coupled to an enzyme assay, yielding a product that can be measured calorimetrically.

Another attractive alternative for detection is the chemiluminescent detection. A variety of applications such as immunohistochemistry, reporter gene analysis, in situ hybridization, RNA labeling and detection, quantitative PCR, DNA sequence analysis, and Southern and Northern hybridization techniques have increasingly employed chemiluminescence. Chemiluminescent detection systems rely on the use of the reporter enzymes that catalyze luminescent reactions. The last step in these systems is the detection of the luminescent reaction.

The two enzymes used most commonly as reporter molecules are horseradish peroxidase (HRP) or alkaline phosphatase (AP). Utilizing HRP, the substrate usually consists of luminal and the presence of hydrogen peroxide is required. The peroxidase enzyme liberates the nascent oxygen from the hydrogen peroxide, which converts the luminal into its product. In this reaction, generation of light is a byproduct. With the inclusion of enhancers, the signal produced peaks quickly, where typically the half-life of such a signal is relatively short, generally in the neighborhood of about an hour.

The chemiluminescent detection systems that use AP as reporter enzyme require Lumiphos™ 530 or 1,2 dioxetane substrates.

DIG Labeling and Biotin Labeling

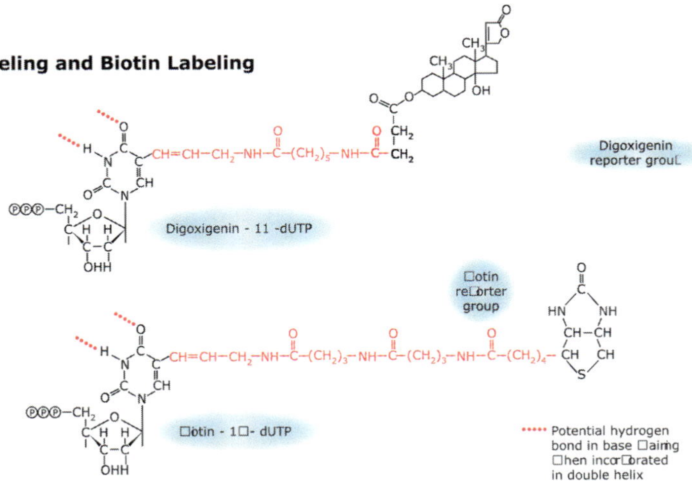

Enzyme Labeling

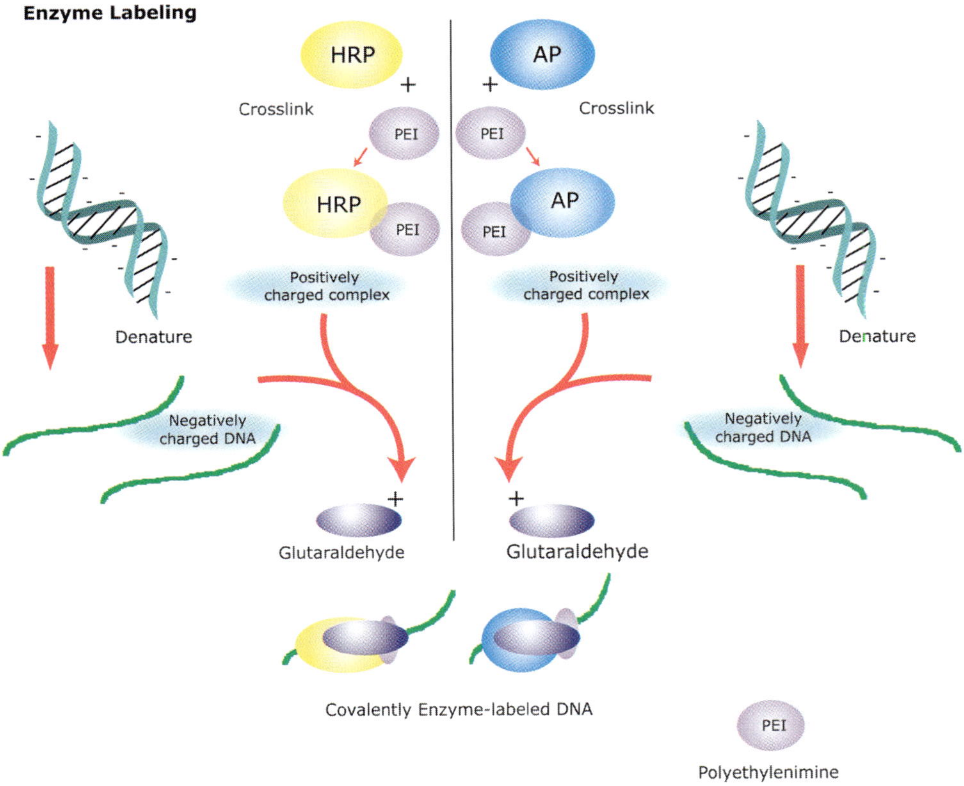

Luminescent reaction in this case is slower than in HRP and the resulting signal remains quite high for a relatively long time. In addition, rejuvenation with fresh substrate may allow ability to re-expose the original membrane for a second signal generation.

Incorporation of haptens into DNA probes is frequently achieved by enzymatic methods such as random primed reactions, nick translation, polymerase chain reaction (PCR), or 3′ end labeling with terminal transferase. In addition, chemical conjugation methods and photocrosslinking methods are also popular. The third variety is specific modification of oligonucleotides at their 5′ ends. The detection of the signal on the membranes proceeds much like Western blot hybridization. The protein binding sites on the membrane are blocked, and probes are detected by specific binding of appropriate secondary conjugates.

Chemiluminescent signals can be detected and recorded using X-ray films. More sophisticated imaging instruments can also be used for visualization of the chemiluminescence signal. If direct detection systems are employed, the DNA probe is labeled with the reporter enzyme itself (HRP- or AP-labeling; see figure). This process eliminates secondary incubation, cutting down the time for detection reactions, and also results in lower backgrounds. However, conjugation of the enzyme to the DNA probe can decrease hybridization effeciency due to steric hindrance afforded by the enzyme molecules. Another factor that has to be taken into account is the activity of the enzyme. Elevated temperatures commonly used in hybridizations and washes have detrimental effects on enzyme activity. Recently, thermostable enzymes, particularly AP, have been used to overcome these difficulties, and conjugation of AP to the 5′ ends of oligonucleotide probes has been achieved. Such 5′ amino-linked oligonucleotides can be used as probes in hybridization subsequent detection.

Recent improvements in the capabilities of nonisotopic labeling systems have led to increased sensitivities. In addition to being a nonhazardous alternative to radioactive labeling, these systems permit labeling of probes and preservation for longer periods of times because of stability of probes.

Suggested Reading

Seibl, R., H.J. Holtke, R. Ruger, A. Meindl, H.G. Zachau, R. Rasshofer, M. Roggendorf, H. Wolf, N. Arnold, and J. Wienberg, et al. 1990. Non-radioactive labeling and detection of nucleic acids. III. Applications of the digoxigenin system. Biol. Chem. Hoppe Seyler *371*:939-951.

Holtke, H.J., R. Seibl, J. Burg, K. Muhlegger, and C. Kessler. 1990. Non-radioactive labeling and detection of nucleic acids. II. Optimization of the digoxigenin system. Biol. Chem. Hoppe Seyler *371*:929-938.

Kessler, C., H.J. Holtke, R. Seibl, J. Burg, and K. Muhlegger. 1990. Non-radioactive labeling and detection of nucleic acids. I. A novel DNA labeling and detection system based on digoxigenin: anti-digoxigenin ELISA principle (digoxigenin system). Biol. Chem. Hoppe Seyler *371*:917-927.

Linked Topics:
Cis-regulatory elements
Embryonic stem cells
Transcription factors
Transgenic animals

Physical Assault Delivery Methods

Electroporation: A short pulse of electrical charge transports foreign DNA into the target cells, probably through electrophoresis of DNA into cells and temporary pore formation in the cell membrane. The recipient cells are placed in special electroporation cuvettes for this manipulation. The DNA to be transported may contain a reporter gene, which permits confirmation of the transfer via reporter gene assay. Although the foreign gene is expressed only transiently, such transient expression can be useful in measuring the activities of related gene constructs (to locate *cis* regulatory elements like promoters, transcription factor binding sites, etc.). Since target cells as diverse as human, animal, yeast, and bacterial cells are all amenable to DNA delivery by electroporation, this method is the most versatile of the currently available ones.

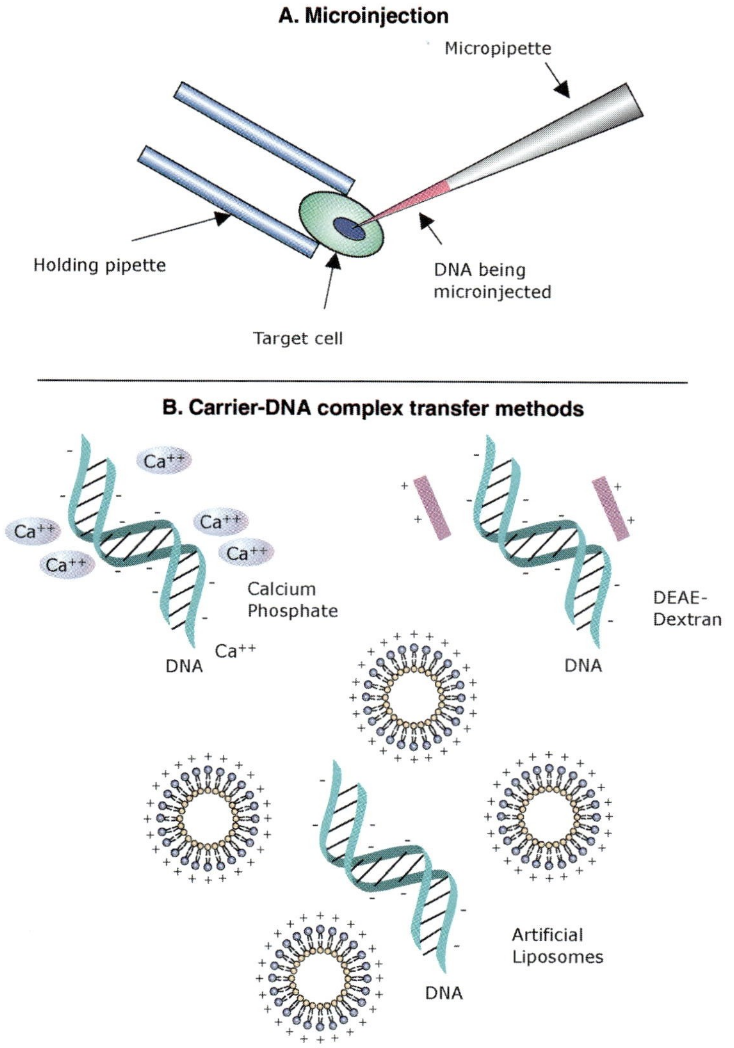

Microinjection: While holding the target cell in place at the tip of a blunt micropipet by gentle suction, a very fine-tipped pipet is used to insert DNA into the cytoplasm or nucleus. Even though the process is tedious, microinjection of genes into embryonic stem cells is a major route for the production of transgenic animals (Figure A).

Microprojectile-mediated transfer: DNA-coated gold or tungston microprojectiles can be propelled to invade a target cell following acceleration by pressurized discharge. The microprojectiles traverse into the target tissue, the depth of tissue delivery being controlled by helium pressure. The DNA introduced by this method can be expressed both transiently and stably. An interesting application of this technology is seen in genetic immunization by propelling genes for antigens into an animal's skin.

Carrier/DNA Complex Transfer Methods

Calcium phosphate microprecipitation-mediated transfer: Desired DNA sequence dissolved in phosphate buffer is added slowly to a solution of calcium chloride. The calcium phosphate precipitate formed traps adsorbed DNA, which is internalized by the target cells. Transfection with microprecipitates is an inexpensive choice for integrating a gene into cultured mammalian cells or embryonic stem cells (Figure B).

Liposome-mediated transfection: When polycationic lipid molecules are hydrated with uncharged lipid molecules in aqueous solution containing DNA, unilamellar lipid-bilayer vesicles (liposomes) are formed, incorporating DNA. The negative charge on DNA phosphodiester backbone is neutralized by the lipids. The overall positively charged carrier liposomes can be transferred easily into the cell due to the net negative charge on cell membrane. Additionally, membrane fusion and endocytosis may also be mechanisms of liposome transfer. Attachment of specific surface receptor ligand to liposome permits receptor-mediated endocytosis. However, the efficiency of gene transfer via liposomes is low, because most of the internalized DNA is degraded due to uptake into lysosomes. This nonviral alternative, which allows both stable and transient expression, is most commonly used for gene transfer in cultured cells.

Diethylaminoethyl (DEAE)-dextran-mediated DNA transfer: Sequential interaction between positively charged DEAE-dextran, negatively charged DNA, and negatively charged plasma membrane results into cellular uptake by endocytosis. Besides its use for only transient expression, toxicity to cells due to high concentrations of DEAE-dextran can be a limitation.

Suggested Reading

Davis, M.E. 2002. Non-viral gene delivery systems. Curr. Opin. Biotechnol. *13*:128-131.

Lechardeur, D. and G.L. Lukacs. 2002. Intracellular barriers to non-viral gene transfer. Curr. Gene Ther. *2*:183-194.

Takakura, Y., M. Nishikawa, F. Yamashita, and M. Hashida. 2002. Influence of physicochemical properties on pharmacokinetics of non-viral vectors for gene delivery. J. Drug Target. *10*:99-104.

Duncan, R. 2003. The dawning era of polymer therapeutics. Nat. Rev. Drug Disc. *2*:347-360.

Trezise, A.E. 2002. In vivo DNA electrotransfer. DNA Cell Biol. *21*:869-877.

Vijayanathan, V., T. Thomas, and T.J. Thomas. 2002. DNA nanoparticles and development of DNA delivery vehicles for gene therapy. Biochemistry *41*:14085-14094.

NUCLEIC ACID HYBRIDIZATION

Linked Topics:
Allele-specific oligonucleotides
In situ hybridization
Nonisotopic labeling

Nucleic acid hybridization is a method of using labeled nucleic acid probe molecules (either DNA or RNA) to identify unlabeled target nucleic acid molecules (DNA or RNA) closely related in base sequence. If the probe molecule is double-stranded DNA, separation of the two DNA strands (denaturation or melting) is a prerequisite. The same applies to the target molecule if it is double-stranded. Typically, this is achieved by heating or alkaline treatment.

After mixing the strands of probe with single strands of target DNA or RNA, the strands with complementary base sequence hybridize by annealing or renaturation. The complementary probe DNA strands can reanneal to form homoduplexes as can the complementary target DNA strands. However, it is the heteroduplex formation between the probe DNA strand and the complementary target DNA strand that defines the usefulness of the hybridization assay. Usually the probe consists of a homogeneous solution of a specific small sequence, while the target is a heterogeneous mixture of many different sequences. A number of factors, such as strand length, base composition, chemical environment, etc., determine the energy required to separate two perfectly complementary DNA strands. Longer homoduplexes containing a large number of hydrogen bonds (such as high GC content DNA molecules) require more energy to separate them. Typical radioisotopic or nonisotopic labeling procedures result in relatively short DNA probes (<500 bp), reducing the effect due to strand length. While the presence of monovalent cations like sodium stabilizes the duplexes, higher pH or chemical denaturants, such as formamide and urea, destabilize the duplexes through chemical disruption of the hydrogen bonds. Another aspect affecting hybridization of the nucleic acid duplex stability is temperature. Melting temperature (T_m) corresponds to the midpoint of temperature range over which transition is observed from the fully double-stranded form to the completely single-stranded form. This transition can be monitored for the target DNA by measuring the change in optical density of DNA at 260 nm in the UV range. Since the UV absorption by double-stranded DNA is less than that of free nucleotides, as the double-stranded DNA is gradually heated, there is an increase in absorption at 260 nm (hyperchromicity). Performing this assay prior to hybridization experiments helps define the temperature at which 50% of the double-stranded DNA is denatured. This point is called T_m, and for mammalian genomes with base composition of about 40% GC, the T_m is about 87°C. Thus, four factors—concentration of monovalent cations, pH, effective percentage of denaturants, and temperature—determine the success of hybridization reactions.

Often, hybridization conditions are chosen to be relatively relaxed, keeping the hybridization temperature as much as 25°C below the T_m so as to promote heteroduplex formation. However, in this case, the posthybridization washes for removal of excess unbound probe can be conducted under more stringent conditions so as to disrupt all duplexes except those between closely related sequences. The probe–target heteroduplexes are most stable thermodynamically when the region of duplex formation contains perfect base matching. Enhancing the stability of partially mismatched heteroduplexes may be desirable in some cases, such as identification of comparatively divergent members of a multigene family or of interspecies relatedness of homologous genes. In such cases, concentration of sodium chloride is increased and temperature is reduced, thereby reducing the hybridization stringency. Base complementarity over more than 100 bp is often sufficient to stabilize heteroduplexes. Conversely, hybridization stringency can be increased by lowering the concentration of sodium chloride and increasing the temperature. This encourages dissociation of mismatched heteroduplexes, which is particularly relevant when oligonucleotide probes (such as allele-specific oligonucleotides) are used to detect single nucleotide polymorphisms (SNPs). Addition of a denaturant, such as formamide, allows a reduction in the hybridization temperature (by 0.72°C for RNA–DNA hybrids for every 1% additional formamide). This is very useful for probes that are directly labeled with thermolabile enzymes, such as horseradish peroxidase (as in chemiluminiscent nonisotopic labeling). Variations of the same principles described here are used in the in situ hybridization techniques.

Suggested Reading

Zhang, P., N. Briones, C.G. Liu, C.K. Brush, T. Powdrill, Y. Belosludtsev, and M. Hogan. 2001. Acceleration of nucleic acid hybridization on DNA microarrays driven by pH tunable modifications. Nucleosides Nucleotides Nucleic Acids *20*:1251-1244.

OLIGONUCLEOTIDE SYNTHESIS

Linked Topics:

Fluorescence resonance energy transfer

Gene cloning methodologies

Microarrays

Modifications of PCR

Modified oligonucleotides

PCR and PCR optimization

Site-directed mutagenesis

Southern and Northern hybridization

The in vivo polynucleotide extension processes for synthesis of DNA and RNA are based upon enzymatic addition of nucleotides to the priming sequences in a 5′ to 3′ direction, dictated by the template polynucleotide sequence to which the primer is annealed. The same principle is followed in polymerase chain reaction (PCR) and different modifications of PCR, as also for in vitro transcription using M13 bacteriophage-based transcriptional vectors. In contrast, the synthetic oligonucleotides are created using template-independent chemical polymerization processes wherein, starting with an initial immobilized nucleotide, the subsequent nucleotides are added in a 3′ to 5′ direction. The ability to employ automated synthesizers for producing large quantities of synthetic oligonucleotides has been pivotal in advances in gene cloning methodologies via creation of adapters, linkers, multiple cloning sites, etc. These short defined stretches of DNA or RNA sequences can be used as primers for PCR, as probes for use in Southern or Northern hybridization or gel shift assay, or to introduce sequence modifications in site-directed mutagenesis. Modified oligonucleotides, such as peptide nucleic acids (PNAs), can be prepared by using structurally altered nucleotide analogs for applications in antisense therapeutics. In situ synthesis of probe oligonucleotides in DNA microarrays is another exciting application.

Phosphite tri-ester synthesis is most commonly used chemistry in automated synthesizers. At the beginning of synthesis, the initial nucleotide is attached to the solid support [polymer or controlled-pore glass (**CPG**)] via its 3′ end, while the 5′ end is capped with a dimethoxytrityl (DMT) group. Through solvent-mediated reduction, the DMT group is reduced (**deprotection**), and a new incoming nucleotide is allowed to form a trivalent phosphite tri-ester bond through its 3′ end with the first nucleotide. This process is called **coupling**. A more stable pentavalent phosphate tri-ester bond formation follows via a process of **oxidation** of the phosphite tri-ester intermediate, resulting in a dinucleotide sequence. Sequential addition of DMT then allows **capping** of the 5′-OH ends of the unreacted anchored nucleotides to prevent their participation in further elongation. The dinucleotide is then used as a template for the addition of further nucleotides through the same processes of deprotection, coupling, oxidation, and capping. After the requisite number of cycles and addition of nucleotides, amino groups are removed from adenine, guanine, and cytosine nucleotides, and the final deprotection is carried out, removing the protecting DMT group from the terminal nucleotide and generating a polynucleotide with the 5′-OH end. The synthetic oligonucleotide is then released from the solid support. Although efficiency of each individual synthetic step is maintained at >98%, the automated oligonucleotide synthesis is known to falter and incorporate incorrect nucleotides beyond a certain threshold of length, despite the sophistication of the controlling software. Typically, 100-nucleotide-long oligonucleotides are the limit of synthesis.

This versatile method of oligonucleotide synthesis can be used to allow intrasynthetic nonisotopic labeling [by amino linkage of 5′ labels like biotin or fluorochromes used in fluorescence resonance energy transfer (FRET)] and modification of either the backbone (as in phosphorothioates, methylphosphonates, phosphorodithioates, PNA) or termini (like incorporation of bleomycin) to protect against nuclease attack. The label attachment typically includes a spacer-molecule chain, permitting the use of the synthetic oligonucleotide as a probe for hybridization. Another area of intense development is the synthesis of **branched (dendritic) oligonucleotides**, where branching of monomers is achieved through linkage of one nucleotide to three others by 5′, 3′, and 2′ linkages, each branch being labeled at high density. Such branched oligonucleotides are being used for in vitro diagnostics for a number of infectious diseases.

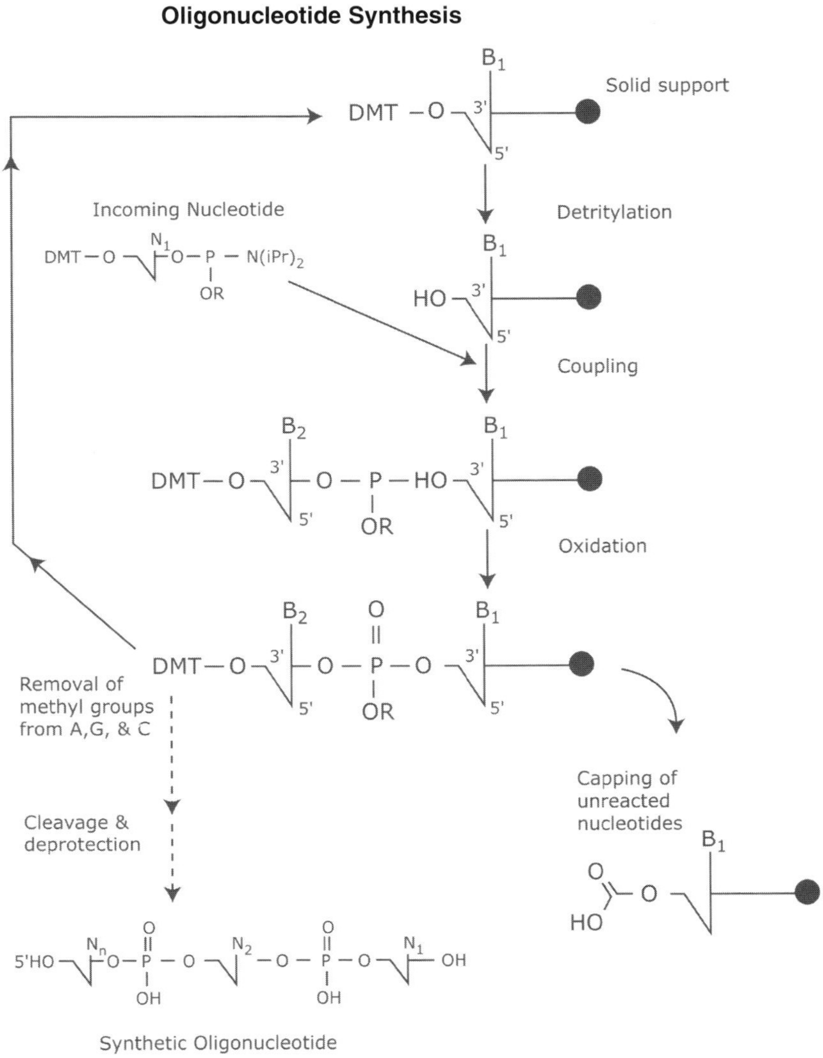

Suggested Reading

Pon, R.T., S. Yu, and Y.S. Sanghvi. 2002. Tandem oligonucleotide synthesis on solid-phase supports for the production of multiple oligonucleotides. J. Org. Chem. *67*:856-864.

Pon, R.T., S. Yu, and Y.S. Sanghvi. 2001. Multiple oligonucleotide synthesis in tandem on solid-phase supports for small and large scale synthesis. Nucleosides Nucleotides Nucleic Acids *20*:985-989.

Reese, C.B. and Q. Song. 1997. A new approach to oligonucleotide synthesis in solution. Nucleic Acids Symp. Ser. *37*:1-2.

Pfleiderer, W., S. Matysiak, F. Bergmann, and R. Schnell. 1996. Recent progress in oligonucleotide synthesis. Acta Biochim. Pol. *43*:37-44.

p53

Linked Topics:
Apoptosis
Cell cycle
Genetic therapy
Molecular chaperones

Discovered in 1979 by several groups, the 53-kDa protein p53 was originally considered to be a tumor antigen based on characteristic nuclear accumulation in tumor cells. The early p53 cDNA constructs derived from tumor cells efficiently induced neoplastic transformation of primary rodent cells when co-transfected with viral oncogenes, such as *ras* or *E1a*, which led to the notion that p53 was an oncogene. It was not until 1987 that loss of heterozygosity on the chromosomal region where the p53 gene resides (17p13.1) was noted in human cancers. It was also shown that not only did the wild-type p53 gene isolated from normal cells fail to induce neoplastic transformation, but it also inhibited both tumor cell growth and neoplastic transformation of primary rodent cells. These findings absolved p53 and reclassified it as a tumor suppressor gene.

The p53 tumor suppressor gene product is now recognized as a vital transcription factor and a member of the DNA damage-response pathway, which is involved in the maintenance of genomic integrity. Loss of p53 leads to increased genomic instability characterized by an increased frequency of gene amplifications, chromosomal rearrangements, and mutations. The p53 gene is frequently mutated in a majority of human malignancies, suggesting that p53 may serve as a common denominator for controlling growth of different kinds of cancers despite their biologic heterogeneity. The p53 gene has been found to be predominantly mutated in families with Le Fraumeni syndrome, in which the family members suffer multiple primary tumors typically including soft tissue sarcoma, osteosarcomas, tumors of the breast, brain, and adrenal cortex. The mutations in these tumors tend to be in the third exon or the third intron, while somatic mutations can be anywhere.

Composed of 393 amino acid residues, p53 protein contains an acidic transcription activation domain (residues 20–42), the sequence-specific DNA binding domain (residues 100–300), the nuclear localization signal domain (residues 316–325), and the oligomerization domain (residues 319–360). While the oligomerization domain is shown to be necessary for transcription activation by p53, it is the DNA-binding domain that displays most of the missense mutations (mutational hot spots).

Several genes crucial in the cell cycle and apoptosis have been identified to be transcriptionally activated by p53. DNA damage by ionizing radiation, UV, or DNA-damaging agents causes a rapid accumulation of p53 proteins in the nucleus of the cell in G_1 phase. The p53 protein, in turn, activates downstream genes whose products may be involved in tumor suppression, such as $p21^{waf1/cip1}$ (G_1/S checkpoint hypothesis). p53 also regulates the G_2/M checkpoint of the cell cycle by inhibiting transcription of cyclin B1 and cdc2. p53 also acts in the transcription-independent pathway by way of protein–protein interaction. Thus, the protein MDM2 (murine double minutes) binds to both wild-type and mutant forms of p53, constituting a negative feedback pathway to regulate p53 function. Several transcriptional factors bind p53 and together repress transcription of proliferation-related genes. On the other hand, heat shock protein 70 (hsp70), one of the molecular chaperones, preferentially binds to the mutant form of p53. p53 also plays a direct role in DNA replication, nucleotide excision repair, and homologous recombination pathway. Insights into the manifold functions of p53 have provided for the prospects of multimodality cancer therapy, including chemotherapy, immunotherapy, and genetic therapy. Termed "molecule of the year" by the journal *Science* in March 1994, p53 is considered the **guardian of the genome.**

Suggested Reading

Haupt, S., I. Louria-Hayon, and Y. Haupt. 2003. p53 licensed to kill? Operating the assassin. J. Cell. Biochem. *88*:76-82.

Fojo, T. 2002. p53 as a therapeutic target: unresolved issues on the road to cancer therapy targeting mutant p53. Drug Resist. Updates *5*:209-216.

Sherr, C.J. and F. McCormick. 2002. The RB and p53 pathways in cancer. Cancer Cell *2*:103-112.

PCR AND PCR OPTIMIZATION

Linked Topics:
Modified oligonucleotides
Oligonucleotide synthesis

Developed by Kary Banks Mullis in 1983, the polymerase chain reaction, known as PCR, has emerged as one of the most elegant and powerful tools in molecular biological applications. PCR allows amplification of even a minute amount of DNA segments using a thermostable DNA polymerase such as *Taq* DNA polymerase, two typically 20–25 nucleotide oligonucleotides (primers)—one complementary to the (+) strand at one end of the target sequence to be amplified and the other complementary to the (-) strand at the other end, and a pool of nucleotide triphosphates. PCR proceeds through 25–40 repetitive cycles of template DNA denaturation, primer annealing, and extension of newly formed polynucleotide chains. The theoretical maximum number of copies achieved is approximated by 2^n, where n is the number of cycles performed. Owing to its ability to yield millions of molecules from even one starting molecule of DNA, PCR has now been firmly anchored in the armament for clinical diagnosis and forensic analysis. Numerous modifications of PCR have been developed, such as reverse transcription PCR (RT-PCR), touchdown PCR, quantitative PCR, real-time PCR, and others. The PCR method made the Human Genome Project possible by effectively replacing laborious bacteria-based gene cloning methodologies. Mullis was awarded the 1993 Nobel Prize in Chemistry for the invention of PCR.

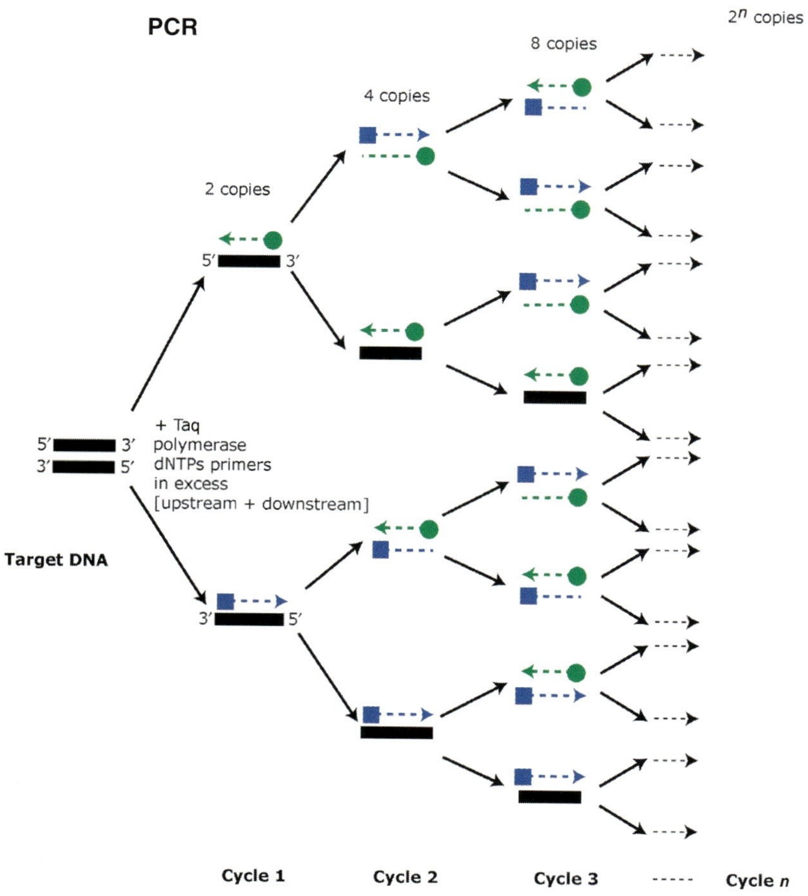

Two major classes of variables in PCR are reaction components (e.g., template, buffer, DNA polymerase and primers) and thermal cycling (e.g., denaturation, annealing, and extension).
Depending upon the objective for PCR (detection, analysis, or preparation of a target), the reaction needs to be optimized for one or more of the above variables.
Except for preparative PCR, for most other applications it is often enough to limit the optimization process primarily to arriving at the correct annealing temperature, and later to fine-tune reaction concentration of Mg++ ions, primer, and template DNA.

Hot-start PCR: During sample preparation at room temperature, complexes of nonspecific primer-template may be generated due to partial sequence homology between the two. Since the thermostable DNA polymerases function over a wide temperature range, starting as low as 37°C, they tend to extend primer sequences in these partial hybrids, resulting into nonspecific amplification of incorrect amplicons earlier on during PCR. By denaturing the misannealed partial hybrids prior to thermal cycling (hot start), extension of these nonspecific complexes can be prevented. Hot-start PCR is thus a common and easy protocol to improve yield and increase specificity. There are a number of variations in methodology to achieve a "hot start." The basic principle is that a key component necessary for amplification (e.g., primers, polymerase, Mg++ ions, or dNTPs) is withheld from the reaction mix at permissive temperatures, until the reaction reaches a temperature above the optimal annealing temperature of the primers. This sequestration is most conveniently done by trapping the component in wax beads, such that the reaction has all the necessary components only after wax melts at above 80°C. An alternative common approach is to add the polymerase in complex with its antibody, the enzyme being released by denaturation of the complex only at higher temperature. The hot-start thus minimizes the spurious side reactions associated with low temperature extension of misprimed oligonucleotides and a more sequence-specific amplification occurs preferentially.

Gradient PCR: Inappropriate annealing temperature results in mispriming and PCR artifacts. The annealing temperature can be optimized in an experiment where a linear temperature gradient of up to 8 or 12 different annealing temperatures (typically from 48°C-60°C) is created into the tube-holders of a gradient thermal cycler. Once an optimal annealing temperature is determined, which yields a strong target amplicon, subsequent experiments are carried out with the heating block programmed for a single-temperature annealing mode.

Nested PCR: When amplification of target gene sequences must preclude amplification from pseudogenes, a very high degree of specificity is required for PCR. Similar need is obvious when the target DNA is present in extremely low copy number (as in the case of detection of viral nucleic acid or for transcripts from a small number of epithelial cancer cells in blood). An extra level of stringency can be imposed onto PCR in such cases by using an additional pair of primers complementary to internal target sequence ('nested primers'). In order to counter the unfavorable effects of increased sensitivity, the primary primer pair is chosen to have higher Tm than the 'nested' pair. The reaction is thus carried out initially at higher annealing temperatures so as to reinforce the correct amplicon, followed by the consolidation of the smaller nested amplicon at lower annealing temperatures.

Suggested Reading

Birch, D.E. 1996. Simplified hot start PCR. Nature *381*:445-446.

Sybesma, W., J. Hugenholtz, I. Mierau, and M. Kleerebezem. 2001. Improved efficiency and reliability of RT-PCR using tag-extended RT primers and temperature gradient PCR. BioTechniques *31*:466-470.

PHARMACOGENETICS AND PHARMACOGENOMICS

Linked Topics:
Functional genomics
Gene therapy
Genetic testing
Human Genome Project
Microarrays
Single nucleotide polymorphisms

The concept of **pharmacogenetics** involves the use of the data from molecular genetic analyses to predict possible disease development or progression and advising individual patients regarding possible choices for prevention or therapy. It combines the disciplines of biochemistry and pharmacology to correlate phenotypic biomarkers, such as drug-induced toxicity, with genetic characterization. **Pharmcogenomics**, on the other hand, is the use of data (from high-throughput molecular assays like microarray technology from the burgeoning field of genomics and bioinformatics) to determine a molecular profile of an individual (for unique faculties such as responsiveness to different classes of drugs) and to design or develop a personalized drug regimen that would be effective. Thus, while pharmacogenetics will mostly be used by clinical geneticists and experts in the field of preventive social medicine, pharmacogenomics is being developed by biotechnology-based pharmaceutical companies for applications in novel drug syntheses. Pharmacogenetics deals with an individual's response to certain drugs, and pharmacogenomics addresses the potential of genomic technology for commercial application in drug development and therapy. Genetic makeup of individual patients is now considered a potential determinant of efficacy, metabolism, and transport of drugs in the patient. After its entry into the body, each drug interacts with numerous proteins (such as carrier proteins, transporters, metabolizing enzymes, receptors, and so forth), which determine the absorption, distribution, excretion, and target site of action for the drug as well as the pharmacological response. The downstream molecular events may also vary among individuals. Since genomics is giving us clues to understand the molecular basis of this variability, pharmacogenomics is increasingly replacing the candidate gene approach of classical genetic studies.

Pharmacogenetics aims to elucidate the relationship between inter-individual differences in response to drugs and the genetic variability affecting drug metabolism or drug targets. The ultimate goal of these studies is to predict drug response at the level of the individual patients. It is becoming evident that large-scale single nucleotide polymorphism (SNP) detection will play a critical role in this process. Genetic testing to identify patients to whom a particular drug can be administered safely and effectively may offer the pharmaceutical products a competitive advantage.

The rapidly changing nature of functional genomics and proteomics technologies makes them untenable for integration into large pharmaceutical companies. This has created a niche for platform pharmacogenomics companies that provide their expertise in genomics and proteomics to the pharmaceutical industry in order to effect a scalable drug discovery and development. The collaboration between pharmaceutical and biotechnology companies, therefore, is on the rise.

These approaches are still in the very early stages of development and will have to deal with ethical issues and cost before they become operationally commonplace. Nevertheless, falling somewhat midway between clearly feasible and proven genetic screening and the yet-to-be-validated gene therapy approaches, the fields of pharmacogenetics and pharmacogenomics do represent highly useful products of the Human Genome Project and the fledgling field of proteomics.

Suggested Reading

Pagliarulo, V., R.H. Datar, and R.J. Cote. 2002. Role of genetic and expression profiling in pharmacogenomics: the changing face of patient management. Curr. Issues Mol. Biol. *4*:101-110.

PHYSICAL MAPPING

Linked Topics:

Alu repeat sequence

Bacterial artificial chromosomes

Bacteriophage λ

In situ hybridization

Modifications of fluorescence in situ hybridization

Nonisotopic labeling and detection

P1 bacteriophage and PAC vectors

Pulsed field gel electrophoresis

Restriction endonuclease

Yeast artificial chromosomes

While the genetic map, a step prior to the Human Genome Project, permitted knowledge of the positional interrelationships of the genes or polymorphisms based on meiotic recombination frequencies, the physical map provided a set of ordered DNA clones and hence, molecularly characterized and localized DNA sequences along the chromosomes. Whereas a genetic map determines a chromosomal location of trait-related genes, a physical map helps in isolating these genes. Starting with a restriction endonuclease cleavage map of a large region, the physical mapping proceeds to determine the sequence of nucleotides at specific loci.

Historically, the first physical map of the human genome was provided by Giemsa banding techniques that distinguished different portions of chromosomes. Although the resolution of this map was low (as much as 10 Mb of DNA), it provided a scaffold for ordering the locations of human DNA sequences by chromosome in situ hybridization or fluorescense in situ hybridization (FISH), which improved the resolution to discern 2–5 Mb. Modifications of FISH, like interphase FISH, improved this resolution even further (1 to 0.1 Mb). High-resolution physical mapping was the next advance with the arrival of first bacteriophage λ or P1 bacteriophage and cosmid, and then yeast artificial chromosome (YAC), bacterial artificial chromosome (BAC), and P1-derived artificial chromosome (PAC) cloning systems. Clones with overlapping DNA segments are retrieved and ordered into contiguous stretches (contigs). YAC clones can carry as much as 1-Mb inserts, whereas the cosmids carry up to 40-kb-long inserts. The YAC clones thus reduce the number of contigs to be ordered. Finally, the resolution was further increased stepwise by using sequence-tagged sites (STS) (about a hundred kbp), expressed sequence tags (ESTs) (around 40 kbp), down to 1 bp by DNA sequencing. The high-throughput automated DNA sequencing has immensely accelerated this "last leg" of the Human Genome Project.

The large-insert contigs from PAC, BAC, and YAC libraries are assembled by either Alu-polymerase chain reaction (PCR) hybridization, DNA fingerprinting, or STS-content mapping, while the small insert contigs from λ and P1 bacteriophage and cosmid clones are ordered by DNA fingerprinting alone. For the **Alu-PCR hybridization** in case of YAC clones, for example, Alu repeat sequence primers are used to perform PCR for different YAC clones, and labeled PCR products of one clone are used to hybridize with unlabeled PCR products of the other clones in the panel. The positive cross-hybridizing clones are further subjected to restriction endonuclease cleavage mapping using rare cutter restriction endonucleases such as *Not*I, *Sfi*I, *Bss*HI, and others, in pulsed field gel electrophoresis (PFGE). The shared cleavage sites are used to align the sequences into contigs. **DNA fingerprinting** for YAC clones is carried out by restriction endonuclease digestion of DNA, followed by electrophoresis, blotting, and hybridization with labeled probes for either L1 repeats or transposable human elements (THE), which yields a fingerprint pattern. Such patterns are compared for different clones, and the clones that share the patterns are further mapped by restriction endonuclease cleavage sites and ordered into contigs. The increasing availability of numerous STS sequences in public databases and of the high-throughput PCR assays has allowed **STS content mapping** to become the most favored method for contig ordering for YACs. In the case of DNA fingerprinting for cosmids, following restriction endonuclease cleavage, the digested fragments are labeled and detected by the appropriate nonisotopic detection system, such as chemiluminiscence. Alternatively, the label is introduced at the ends of the fragments via a fluorochrome-labeled linker, and the fragments are detected by an automated fluorescence scanning during electrophoresis. Both these methods thus circumvent hybridization.

Suggested Reading

Le Hellard, S., C.A. Semple, S.W. Morris, D.J. Porteous, and K.L. Evans. 2001. Physical mapping: integrating computational and molecular genetic data. Ann. Hum. Genet. *65*:221-228.

PLASMIDS

Linked Topics:

Bacterial artificial chromosome vectors

Gene cloning methodologies

Human Genome Project

P1 bacteriophages and PAC vectors

Yeast artificial chromosome vectors

Plasmids are circular covalently closed double-stranded DNA molecules that naturally exist extra-chromosomally in bacterial or yeast cells. These molecules are smaller compared to the host genome, ranging from 1 kb to a little more than 200 kb. They carry only a few genes—some of which may be of help to the host cells, such as genes for conferring resistance to antibiotics or those responsible for production of antibiotics, enterotoxins, colcines, restriction modification enzymes, etc. The plasmids replicate independently of the host cell genome, but may require host proteins for their replication and transcription. They can be inherited vertically down the host cell generations, but may also be transmitted horizontally by bacterial conjugation. (The latter is the reason why antibiotic resistance sometimes spreads so rapidly and alarmingly.)

Plasmids have become workhorses of recombinant DNA technology, wherein they are used as vectors to carry and amplify desired foreign DNA molecules. Most often, a plasmid contains one origin of replication (ori), which together with the *cis*-acting control element, constitutes a genetic unit called a replicon. For replication, plasmids carrying replicons such as ColE1 or pMB1 depend not on plasmid-encoded proteins but solely on long-lived host proteins. Such plasmids can, therefore, replicate to a high copy number even after protein synthesis is arrested by antibiotics, such as chloramphenicol or spectinomycin, and are considered to be under **relaxed control** of replication. They can be replicated to up to 3000–4000 copies/host cell. On the other hand, plasmids that carry replicons, such as pSC101, depend on their own replicative protein machinery. Hence, inhibition of protein synthesis cannot bring about increase in copy number of these plasmids. Such plasmids under **stringent control** of replication can exist at only 5 to 6 copies/host cell. For obvious reasons, the gene cloning methodologies requiring high amplification of chimeric plasmid molecules chiefly employ plasmids under relaxed control as vectors. But in certain applications, where very high number of plasmids is deleterious or lethal to the host cells, the plasmids with stringent control may be preferred.

Mobilization of a plasmid molecule from a host cell to the neighboring cell is dependent on proper functioning of *mob* gene in plasmid DNA. In vectors used for gene cloning, this *mob* gene is mutated, and the plasmids are thus rendered incapable of horizontal transmission.

The chimeric plasmid molecules can be introduced into the host cells by a variety of techniques. These host cells are referred to as transformed cells. It is essential to have an ability to distinguish the transformed cells that carry the chimeric plasmids from those that do not. The facility of plasmids to carry genes that confer resistance to antibiotics, such as ampicillin, tetracycline, chloramphenicol, neomycin, canamycin, etc., has been utilized to provide selectable markers during gene cloning.

Natural plasmid molecules can be adapted for use in gene cloning by introducing a variety of restriction endonuclease cleavage sites via a syn-

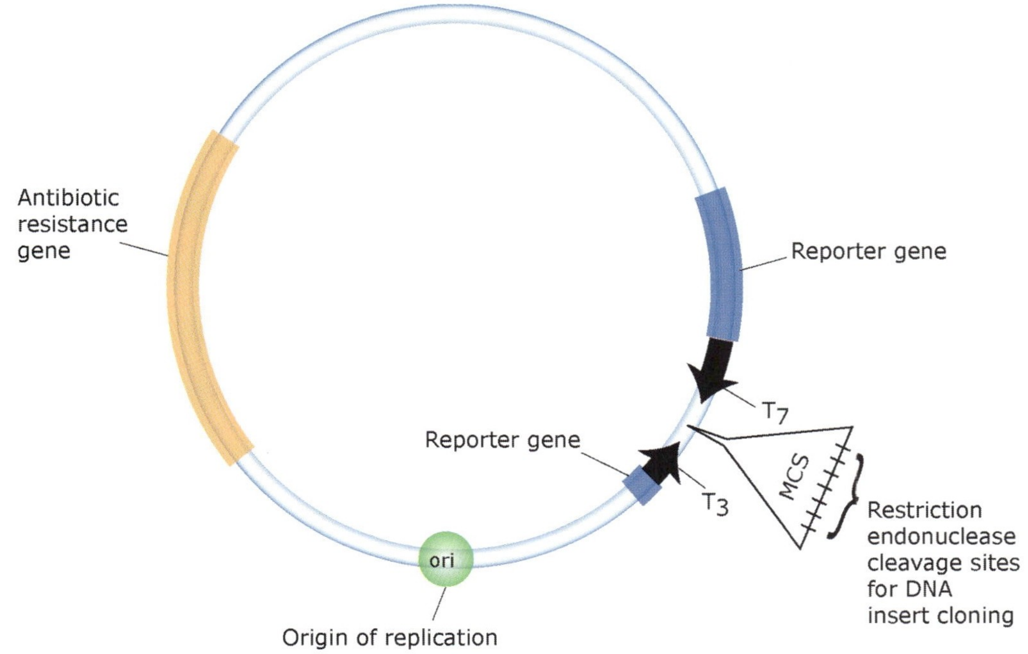

MCS=Multiple Cloning Site
T3=T3 Promotor
T7=T7 Promotor

thetic DNA sequence called **polylinker** or **multiple cloning site** (**MCS**) within a selectable marker gene. One or more of these MCSs can then be used to insert the desired foreign DNA sequence, resulting in a chimeric DNA molecule. pBR322, pUC18, pUC19, pUC118, pUC119, pSP64, pGEM®3, and Bluescript® are some of the examples of plasmid vectors.

Suggested Reading

Dimopoulou, I.D., M.Z. Zaini, and D.W. Crook. 2003. Characterization of plasmids. Methods Mol. Med. *71*:93-99.

Vivian, A., J. Murillo, and R.W. Jackson. 2001. The roles of plasmids in phytopathogenic bacteria: mobile arsenals. Microbiology *147*:763-780.

PULSED FIELD GEL ELECTROPHORESIS

Linked Topics:

Bacterial artificial chromosomes

Gene cloning methodologies

Human Genome Project

P1-derived artificial chromosomes

Southern and Northern hybridization

Yeast artificial chromosomes

Since the matrix structure in the conventional agarose gel electrophoresis limits resolution of DNA fragments to <20 kb, an alternative technique was developed to separate larger genomic fragments derived from manipulation of chromosomal DNA cleaved with rare cutter restriction endonucleases. This technique, termed pulsed field gel electrophoresis (PFGE), is based on the property of large DNA fragments to migrate through a gel matrix to effect a slow turn around in response to changing directions of electrical field. By positioning the pairs of electrodes at angles to each other, the resultant linear motion of the chromosomal DNA is achieved. The large pieces of DNA thus separated are inserted into artificial chromosome vectors, such as bacterial artificial chromosomes (BACs), yeast artificial chromosomes (YACs), and P1-derived artificial chromosomes (PACs) for gene cloning. Two main variations of PFGE exist: field inversion gel electrophoresis (FIGE) and alternating angle gel electrophoresis. The most commonly used clamped homogeneous electric field (CHEF) gel electrophoresis is a modified form of the alternating angle gel electrophoresis.

Sample preparation for chromosomal DNA: The procedures otherwise used for DNA extraction for smaller molecules are not suitable for large chromosomal DNA due to substantial shearing forces generated during pipetting. Hence, a milder method to isolate DNA is employed, where cells are embedded in agarose blocks before subjecting them to further manipulations. The immobilized cells are treated with gentle lysis buffers containing proteinase K, which is subsequently inactivated by using a suitable inhibitor, and the chromosomal DNA is then cleaved into large pieces by incubating the agarose block with a rare cutter restriction endonuclease. After digestion, the agarose blocks are inserted into gel wells, sealed in place with agarose, and subjected to electrophoresis.

Field inversion gel electrophoresis: In the FIGE system, using the traditional gel apparatus, the polarity of electrical field is reversed repeatedly after specified periods. Due to the difference in size and conformation, the larger molecules take longer to turn around in any direction, not unlike big ships compared with smaller boats in the sea, and are therefore separated from the smaller ones. The time allowed for forward mobility is sufficiently longer than that for reverse, which ensures that the overall resultant motion is forward. The gel electrophoresis is routinely done overnight to complete the separation.

Alternating angle gel electrophoresis: An earlier version of this type of PFGE involved alternating electrical fields, by switching the field angles by 90° back and forth, which disentangles a mixture of large DNA molecules from each other and separates them based on their conformation and size. The shortcoming of this right-angle-polarity-switching technique is that uneven gel migration results in a fan-like pattern. A method for creating a more homogeneous field for uniform separation of large DNA molecules was therefore developed (i.e., contour-clamped homogeneous electric field electrophoresis), which is based on placement of electrode pairs at an angle of 120° from each other. This is typically done by using a hexagonal gel box to separate DNA in a rectangular gel, where voltage direction is altered by 120° every 90 seconds for as many as 24 hours.

PFGE can also be used for Southern hybridization for sequence characterization and significantly aided the progress of the Human Genome Project.

Suggested Reading

Maule, J. 1998. Pulsed-field gel electrophoresis. Mol. Biotechnol. *9*:107-126.

Lahti, C.J. 1996. Pulsed field gel electrophoresis in the clinical microbiology laboratory. J. Clin. Lab. Anal. *10*:326-330.

Bignell, G.R. and I.H. Evans. 1996. Chromosomal localization of genes through pulsed-field gel electrophoresis techniques. Methods Mol. Biol. *53*:79-88.

QUANTITATIVE REVERSE TRANSCRIPTION PCR

Linked Topics:

Microarrays

PCR and PCR optimization

Real-time quantitative PCR

Reverse transcription polymerase chain reaction

Ribonuclease protection assay

Southern and Northern blotting and hybridization

Conventionally, steady-state levels of individual mRNA have been measured by Northern blotting and ribonuclease protection assay. These classically useful techniques, however, are limited by their sensitivity, are labor-intensive, and require a radioisotopic or fluorescent label. Thus, for example, approximately 500,000 copies of a transcript (or at least 0.2 μg of total RNA) are required for detection by ribonuclease protection assay. In addition, this assay can be used for transcript quantification only if the labeled DNA probe is used in molar excess. Since clinical studies often have only small amounts of tissue sample, reverse transcription polymerase chain reaction (RT-PCR) has emerged in the past decade as the most favored method. It is significantly more sensitive, requiring as little as 0.01 ng of total RNA, and is capable of detecting moderately expressed transcripts from even a single cell. It has been modified for quantification, taking into account variation in the number of target molecules and amplification efficiency.

Since initial approaches such as intra-PCR DNA labeling followed by Southern hybridization were at best semiquantitative and laborious, quantification of RT-PCR was sought by including housekeeping gene transcripts [e.g., glyceraldehyde-3-phosphate dehydrogenase (GAPDH), β-actin, cyclophilin, and histone H3] as internal controls. But, since very few genes are expressed in a really constitutive manner across different tissues, this approach was soon replaced by competitive or noncompetitive PCR quantification methods.

Target-related synthetic controls called **mimics** have been developed for exogenous addition in RT-PCR, which are either complementary DNA (cDNA) or complementary RNA (cRNA) obtained by in vitro transcription. These mimics share the primer-binding sequence with the target, but differ from the target sequence such that they are about a 100 bp shorter or longer due to deletion or insertion in the target cDNA (and may contain a novel restriction endonuclease cleavage site). The amplicon size difference allows co-amplification yet clear distinction in same agarose gel.

In **noncompetitive RT-PCR**, the mimic and target mRNA are co-amplified under noncompeting conditions. This assay can provide information about the equivalence point in terms of a linear graph and is able to detect small changes in the level of gene expression. However, the absolute number of target molecules in a sample significantly impacts the number of cycles needed to detect the product in mid-exponential phase. Although cRNA mimics control for the variability in efficiencies of both RT and PCR phases, cDNA mimics (**competitors**) are used more often in **competitive RT-PCR**. Since homologous mimics tend to form heteroduplexes with the target during amplification, heterologous competitors may be preferred, which are prepared using **composite primers** for amplification (wherein a short primer sequence for an unrelated cellular gene is synthesized as a 3′ continuation to one for the target gene sequence). The co-amplified serially diluted competitor and the target cDNA compete with each other for enzyme and primers, allowing determination of an equivalence point (which denotes the amount of the target mRNA) via a log scale plot. Besides gel electrophoresis, the alternative approaches for assaying the products of both noncompetitive and competitive RT-PCR include capillary electrophoresis and denaturing high-performance liquid chromatography (dHPLC).

Recent advances in PCR quantification have come about due to elucidation of fluorescence quenching phenomena, such as fluorescence resonance energy transfer (FRET), which have resulted in the development of real-time quantitative PCR. Another quantitative approach is cDNA microarrays.

Suggested Reading

Joyce, C. 2002. Quantitative RT-PCR. A review of current methodologies. Methods Mol. Biol. *193*:83-92.

Freeman, W.M., S.J. Walker, and K.E. Vrana. 1999. Quantitative RT-PCR: pitfalls and potential. BioTechniques *26*:112-125.

RAPID AMPLIFICATION OF cDNA ENDS BY PCR

Linked Topics:

cDNA libraries

Differential display reverse transcription PCR

Gene cloning methodologies

Genomic DNA

Modifications of PCR

Oligonucleotide synthesis

PCR and PCR optimization

Rapid amplification of cDNA ends (RACE) is a polymerase chain reaction (PCR)-based cloning method for the genes expressed at low abundance, which, as the name indicates, allows researchers to obtain clones representing full-length transcripts.

Traditional avenues for cDNA production and representation, like construction of cDNA libraries and identification of differentially expressed mRNAs by differential display reverse transcription PCR (DDRT-PCR), usually result in production of only partial-length cDNA fragments and may take anywhere from days to weeks. The incomplete reverse transcripts are often seen in oligo(dT)-primed libraries. To circumvent this problem, Martin and Frohman developed the RACE-PCR method, where the first step was to reverse-prime the mRNA to cDNA using a gene-specific primer. This was followed by polyadenylation at the 3′ end of the first strand of cDNA using the enzyme terminal deoxyribonucleotidyl transferase (TdT), and then carrying out PCR with an anchored upstream primer containing oligo(dT) sequence at its 3′ end and the gene-specific primer as the downstream primer. The inadequacies of the initial RACE protocol involved high background noise due to generation of nonspecific PCR products (which would warrant a nested PCR to achieve greater specificity) or inefficient homopolymer tailing by TdT (resulting in incomplete 5′-cDNA ends). Collectively, these problems resulted in decreased fidelity, specificity, and yield of cDNA.

Various modifications have since been introduced to the basic RACE protocol, which minimize the time required by eliminating steps. RNA ligase-mediated RACE (RLM-RACE), ligation-anchored RACE (LA-RACE), and adapter-mediated RACE (AM-RACE) are some of the major ones. The **RLM-RACE-PCR** eliminates use of TdT and employs bacteriophage T4 RNA ligase to attach a defined RNA anchor sequence at the 5′ ends of mRNAs before RT using a gene-specific primer. This way, each resulting cDNA strand in the pool encodes an anchored primer sequence. The PCR is then carried out using gene-specific primer and the anchored primer. A slight variation in **LA-RACE-PCR** exploits the property of the same T4 RNA ligase to use deoxribonucleotides as substrate instead of ribonucleotides. Here, mRNA is reverse transcribed using a gene-specific primer, and an anchor primer sequence is ligated to the 3′ end of the first strand of cDNA using the ligase. PCR then continues using the anchor primer and the gene-specific primer. The RNA ligase, however, has low efficiency, which results in low yields. In **AM-RACE-PCR** (also known as **Marathon RACE-PCR**), the double-stranded cDNA synthesis is carried out in a standard manner, and the same adapter sequence is blunt-end ligated at both ends using a T4 DNA ligase. This allows a unified method for 5′ and 3′ RACE and generates an uncloned cDNA library, from which many genes can be amplified. For a full-length cDNA, separate 5′- and 3′-

RACE reactions are performed using two gene-specific primers to generate overlapping 5′- and 3′-cDNA products that span the complete transcript, which can then be cloned and sequenced.

More recently, two major impediments (incomplete 5′ ends and difficulties in adapter ligation) have also been circumvented using the observation that some reverse transcriptases exhibit a TdT-like activity and add three to five dC residues to the 3′ end of cDNA first strand. A synthetic oligonucleotide, with a stretch of dG annealed to the dC-rich cDNA tail, serves as an extended template for RT. The sequence corresponding to 5′ end of the mRNA is thus captured. Here, total RNA is used to address the issue of insufficient availability of starting high-quality poly(A)+ RNA, and the nonspecificity that could consequently result is solved by touch-down PCR.

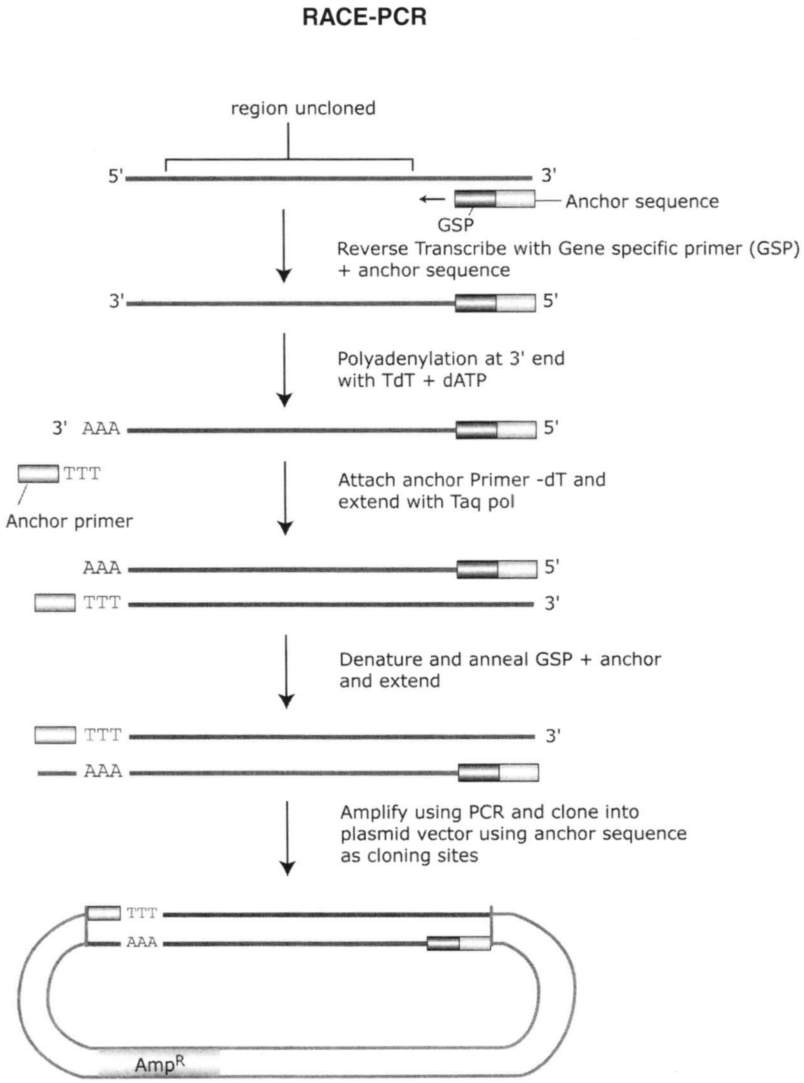

Suggested Reading

Borson, N.D., W.L. Salo, and L.R. Drewes. 1992. A lock-docking oligo(dT) primer for 5′ and 3′ RACE PCR. PCR Methods Appl. *2*:144-148.

Frohman, M.A. and G.R. Martin. 1988. Rapid production of full-length cDNAs from rare transcripts: amplification using a single gene specific oligonucleotide primer. Proc. Natl. Acad. Sci. USA *85*:8998-9002.

Reddy, M.K., S. Nair, and S.K. Sopory. 2002. Global amplification of cDNA from limiting amounts of tissue. An improved method for gene cloning and analysis. Mol. Biotechnol. *22*:223-230.

REAL-TIME PCR

Linked Topics:

Fluorescence resonance energy transfer

Microarrays

Modifications of PCR

Pharmacogenomics and pharmacogenetics

Quantitative PCR

Single nucleotide polymorphisms

TaqMan® and molecular beacons

While the early cycles in polymerase chain reaction (PCR) have undetectable amounts of the DNA product, even the late cycles (plateau phase) yield no quantitative information. The PCR reaches a plateau after exponential amplification, beyond which point, regardless of the starting number of target molecules, the differential quantification is not possible in individual reactions. The quantitative information in a PCR thus emanates from those few (four to five) cycles where DNA molecules grow logarithmically from scarcely detectable to the log-linear phase. Rather than end point analysis of the final amplification product, as in traditional PCR, the real-time PCRs are characterized by the threshold cycle (C_t), beyond which the fluorescence is detectable. The higher the starting number of target molecules, the sooner this threshold is reached. Using the standard curve of samples with a known number of initial molecules, it is possible to determine the initial nucleic acid concentration based on C_t value for the experimental sample.

Compared with other molecular detection techniques, fluorescence offers an advantageous linearity in its response over a large dynamic range. Developments in fluorescence detection technologies, such as SYBR® Green continuous fluorescence monitoring or fluorescent resonance energy transfer (FRET) technologies like TaqMan® and molecular beacons, have brought about a revolution in PCR quantification. Real-time PCR employing one of the above methods tracks the kinetic progress of reaction as the amplification proceeds (in real time), instead of the earlier end point analysis methods. Both these techniques utilize the ability of certain synthetic nucleotide labels to quench the fluorescence of other nucleotide labels. Accumulation of PCR product is measured during the exponential phase of the reaction using fluorescence data, which provides accurate information on initial starting copy number. Variations of this methodology, all of which are automated, are rapidly replacing the more cumbersome competitive and noncompetitive quantitative PCR methods.

Real-time PCR combines both amplification and detection in a single step, thus speeding the process of acquiring experimental data. Using a single tube also eliminates the post-PCR manipulation and reduces the possibility of false positives due to laboratory contamination. Since real-time PCR offers a dynamic range of up to seven orders of magnitude, quantitative measurement is possible for samples containing a few copies of DNA to a few million. Real-time quantitative PCR can quantify nucleic acids for studying gene expression, mutational analysis, disease state, gene dosage, gene knock-out analysis, single nucleotide polymorphism (SNP) screening, and bacterial detection in production cell lines. It can also be used in multiplex PCR for the detection of pathogenic retroviruses. An ability to rapidly carry out expression profiling studies using real-time PCR will substantially boost modern drug discovery based on pharmacogenomics.

Compared to the microarrays used in massively parallel gene expression studies, which (as currently employed) are capable of screening thousands of genes for a small number of samples, real-time PCR assays can screen numerous nucleic acid targets in a finite number of samples, as well as assay numerous samples for a finite number of genes. Besides, while microarrays need up to 1 μg of input RNA, real-time PCR can detect and quantify a target transcript from as little as 10 ng total RNA. Another comparative issue is the detection sensitivity; with a dynamic range of only three to four orders of magnitude, the microarrays may fail to detect a change in expression if it falls outside the optimized detection window.

Suggested Reading

Klein, D. 2002. Quantification using real-time PCR technology: applications and limitations. Trends Mol. Med. *8*:257-260.

Ong, Y.L. and A. Irvine. 2002. Quantitative real-time PCR: a critique of method and practical considerations. Hematology *7*:59-67.

RESTRICTION ENDONUCLEASES

Linked Topics:
Bacteriophages
DNA fingerprinting
Gene cloning methodologies
Physical mapping
Plasmids

In the early 1950s, Luria and colleagues noticed that bacteriophages (or simply "phages") could readily infect and generate plaques against a certain strain of bacteria but were unable to do so against another. The infectivity of a phage, therefore, was "restricted" to one strain of bacteria. In 1962, in an attempt to explain this "restriction" phenomenon, Arber and Dussoix postulated that certain strains of bacteria contained endonucleases that degraded unprotected DNA, while others contained a modification system that protected their own DNA and that of their natural phages from degradation by endonucleases. In the event the phage is unmodified, its DNA is vulnerable to degradation by endonucleases, and such a phage therefore would fail to produce plaques. Understanding of this host-controlled restriction-modification system has resulted in identification, isolation, and purification of perhaps the most important workhorses of molecular biology, the restriction endonucleases, enzymes that cleave double-stranded DNA at or around specific recognition sequences. Smith, Wilcox, and Kelley, working at Johns Hopkins University in 1968, isolated and characterized the first restriction nuclease from *Haemophilus influenzae* bacteria, an enzyme nominated as *Hin*dII. Smith and Nathans were the first to propose a uniform nomenclature system for restriction enzymes. Each of the endonucleases is named according to the following convention: the first capitalized letter corresponds to the first letter of the genus name of host organism used to isolate the enzyme from, followed by the first two letters of the species name. These three letters are italicized. Thus, an enzyme derived from *Escherichia coli* is named *Eco*. A particular source strain from among the multiple strains of a host organism is denoted by identification without italics after the three-letter name (for example, *Eco*R). Finally, should the given strain have more than one endonuclease, a Roman numeral identifies a particular endonuclease (*Eco*RI, *Eco*RIV, etc).

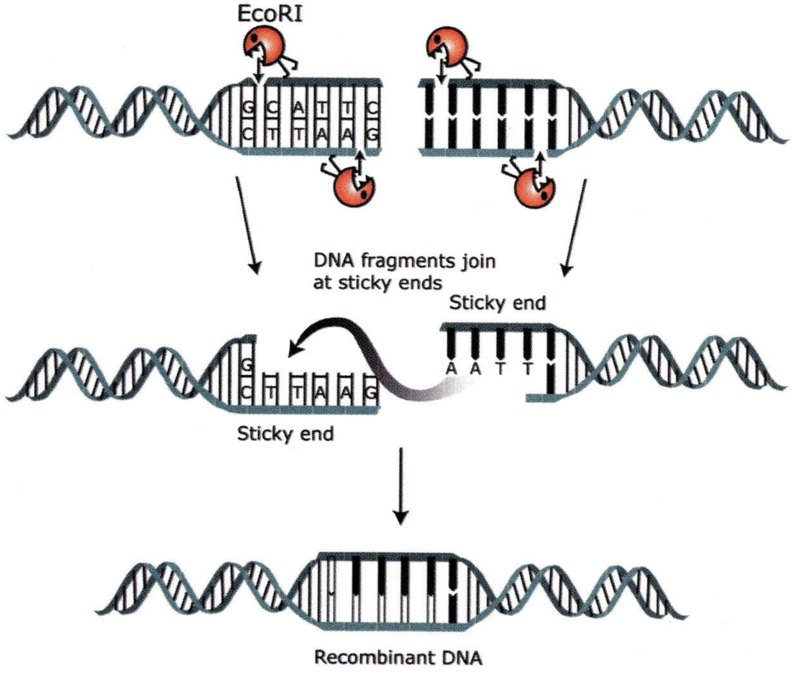

Restriction enzyme systems are divided into three subtypes: type I, I, and III. In type I and III endonucleases, the same molecule harbors both restriction and methylation functions, by virtue of two to three heterologous subunits. A type II endonuclease, however, contains separate restriction and methylation systems, and these are the enzymes widely available and used for studies in molecular biology. While the type I endonucleases cleave 24–26 bp downstream of the recognition site, and the type III endonucleases are even more nonspecific, cleaving as many as 1000 bp away from recognition site; type II endonucleases cleave within a specific 4–8 bp sequence with mirror symmetry around a central axis ("palindromic sequences," like CCGG/GGCC, GGATT/CCTAA, or GAATTC/CTTAAG). Isoschizomers are distinct endonucleases derived from different bacterial strains, which recognize and cleave the same sequence of DNA base pairs. Typically, the type II enzymes cut within the recognition site to leave either a "blunt end," or a 5′or 3′ single-stranded protrusion or overhang, called a "sticky end." The overhangs are usually two to five bases, but one- and three-base overhangs also exist. Longer, GC-rich overhangs ligate most efficiently owing to better stability of the triple hydrogen bonds between the base pairs.

Restriction enzymes optimally function at 37°C and the majority of them can be inactivated with incubation at 65°C for 20 minutes. There are, however, a subset of restriction enzymes, including *Apo* I, *Bcl*I, and *Bsa*HI, that cleave optimally only at a higher temperature. Such enzymes are termed "thermophiles." The number and size of the fragments generated by a restriction enzyme digestion depend on the frequency of occurrence of the cleavage site in the target DNA. Assuming 50% GC content, the shorter the recognition sequence, the more frequent is the occurrence of the restriction site and, therefore, shorter and more numerous are the resulting restriction fragments. Hence, in an application like DNA fingerprinting, DNA is cleaved into multiple small pieces with four-base-cutter restriction enzymes such as *Hae*III. Conversely, DNA is cleaved into larger fragments with eight base cutters for gene cloning methodologies used in physical mapping. Six base cutters are used for cloning into specific regions of plasmids where a series of unique sites known as polylinkers have been engineered.

Suggested Reading

Williams, R.J. 2003. Restriction endonucleases: classification, properties, and applications. Mol. Biotechnol. 23:225-243.

Pingoud, A. and A. Jeltsch. 2001. Structure and function of type II restriction endonucleases. Nucleic Acids Res. *29*:3705-3727.

Williams, R.J. 2001. Restriction endonucleases and their uses. Methods Mol. Biol. *160*:409-429.

Luria, S.E. and M.L. Human. 1952. A nonhereditary, host-induced variation of bacteria viruses. J. Bacteriol. *64*:557.

Arber, W. and D. Dussoix. 1962. J. Mol. Biol. *5*:18.

Dussoix, D. and W. Arber. J. Mol. Biol. *5*:37.

Smith, H.O. and K.W. Wilcox. 1970. A restriction enzyme from *Hemophilus influenzae*. I. Purification and general properties. J. Mol. Biol. *51*:379.

Kelly, T.J., Jr. and H.O. Smith. 1970. A restriction enzyme from *Hemophilus influenzae*. II. J. Mol. Biol. *51*:393.

REVERSE GENETICS: POSITIONAL CLONING

Linked Topics:

Bacterial artificial chromosomes

Exon trapping

Gene cloning methodologies

Genomic DNA and cDNA libraries

Human Genome Project

Nucleic acid hybridization

PCR and PCR optimization

Positional cloning is a location-based gene cloning method. In the traditional functional gene cloning approach (so-called forward genetics), a known biochemical defect (and hence a known protein) is used to identify the corresponding mRNA [either from polysomes or by degenerate polymerase chain reaction (PCR) using primer sequences deduced from amino acid sequence]. The cDNA and subsequently analogous genomic DNA sequence are identified from cDNA and genomic DNA libraries using PCR or nucleic acid hybridization. Alternatively, antibodies for the protein can be used to screen the expression libraries from the cDNA clones. This approach thus follows from disease to function to gene. For the diseases with no known biochemical defect in a protein, however, this forward genetics approach is unsuitable. Positional cloning provides an alternative, where mapped gene position is used to track a gene down, clone it, and then study the effect of mutations on its function.

Given the formidable size of human genome (3×10^9 nucleotides), the magnitude of the positional cloning problem is colossal. Beginning at the level of a subchromosomal region (known either from a translocation or a large deletion or an insertion associated with the disease), a series of corresponding overlapping yeast artificial chromosome (YAC) clones of genomic DNA are used as targets for analysis using polymorphic sequence-tagged site (STS) primers to pinpoint the polymorphism to within 1 cM region (1 centi-Morgan equals approximately 1 million nucleotides), which may still contain as many as 50 genes. This is followed by identification of exon(s) from the contigs, which is a crucial component of positional cloning. Exon identification is achieved using CpG island identification, hybrid selection **zoo blots** (also called multispecies Southern blots or Noah's arc blots), exon trapping, or exon prediction. CpG islands, as detected by *Hpa*II restriction endonuclease digestion, often flank a coding sequence. Similarly, since coding sequences are conserved across species, while intronic sequences diverge evolutionarily, zoo blots can reveal the identity of exons in the sequence used as probe. While the presence of CpG islands and positive hybridization signal in multispecies Southern blots are good flags for the presence of exons in the vicinity, they do not inform about the relatedness of the exons to the target gene. Positional cloning, as employed in the Human Genome Project, proceeded through both the use of commercially available systems for exon trapping as well as exon prediction, by sequencing the contigs using automated sequencers, followed by a search for consensus exon–intron boundaries. Using the exon-containing clone as the probe, cDNA and genomic DNA contigs in the bacterial artificial chromosome (BAC) and cosmid clones are screened for corresponding sequence.

Among the more than fifty genes to be identified by positional cloning are those for Duchenne's muscular dystrophy, cystic fibrosis, kidney disease, colorectal cancer, and breast cancer. While the list of genes cloned by positional cloning is impressive, the methodology is very laborious.

Suggested Reading

Xu, L. and Y. Li. 2000. Positional cloning. Methods Mol. Biol. *136*:285-296.

Boehm, T. 1998. Positional cloning and gene identification. Methods *14*:152-158.

REVERSE TRANSCRIPTION PCR

Linked Topics:

Differential display reverse transcription PCR

In situ hybridization

PCR and PCR optimization

Real-time PCR

The reverse transcription polymerase chain reaction (RT-PCR) method is among the most sensitive techniques available for the analysis of gene expression in vitro. An RNA template is copied into a complementary DNA (cDNA) transcript using the enzyme reverse transcriptase, followed by amplification of the cDNA using PCR. Using this method, it is possible to detect any gene transcript, irrespective of the relative quantity of the specific mRNA, because theoretically, only one mRNA molecule needs to be completely intact between the two primer binding sites.

Three types of reverse transcriptases are commercially available. The mesophilic viral reverse transcriptases are the ones derived from avian myeloblastosis virus (AMV) and Moloney murine leukemia virus (M-muLV), which can reverse transcribe mRNAs up to 10 kb. The third one, rTth, is a thermostable DNA polymerase from *Thermus thermophilus*, which can be used to perform both RT and PCR in a single step and is capable of synthesizing cDNAs in the range of 1 to 2 kb. Besides their optimal reaction temperatures, the important factors to consider while selecting the enzyme are the presence of RNase H activity, which degrades RNA in an RNA–cDNA hybrid, and divalent ion requirements. Three types of sequences can be used to prime RT reactions; oligo$(dT)_{12\text{-}18}$ primers, which bind to the 3′ poly(A) tail present in most eukaryotic mRNAs, random hexanucleotides, which can bind at any complementary sequence throughout the length of mRNA, or gene-specific oligonucleotides. Although the results from a degraded sample may not be quantitative, the relative amount of one RNA molecule can be compared to another RNA molecule in the same sample. For precise determination of the absolute abundance of a specific transcript by RT-PCR, one needs to use of competitive RT-PCR, noncompetitive RT-PCR, or quantitative RT-PCR techniques.

In situ RT-PCR: RT-PCR can be used in situ for the detection of mRNAs that are present at less than 10 copies per cell in fixed cell preparation on slides. The first step, of course, is conversion of cellular mRNA into cDNA using reverse transcriptase. In **direct in situ PCR**, the label (generally digoxigenin or biotin) is incorporated into the amplified product during the PCR step. Digoxigenin, derived from *Digitalis* plants, is not detectable in any other biological material, and, hence, is often preferred over biotin, which is a member of the vitamin B complex. The end result is in the form of enzymatic localization of a chromogen at the site of DNA amplification. Although theoretically simple, direct in situ PCR often tends to yield false-positive results, either due to endogenous mispriming or DNA repair artifact (nonspecific incorporation of labeled nucleotides into fragmented endogenous DNA undergoing repair). Such a false-positive signal is typically seen in the nuclei of apoptotic and senescent cells where DNA fragmentation occurs. Using exonuclease-free DNA polymerase, pretreatment of the cells with T4 ligase to repair the nicks in endogenous DNA, or initial thermal cycling using unlabeled nucleotides are some of the alternatives

RT-PCR primed by Oligo-dT

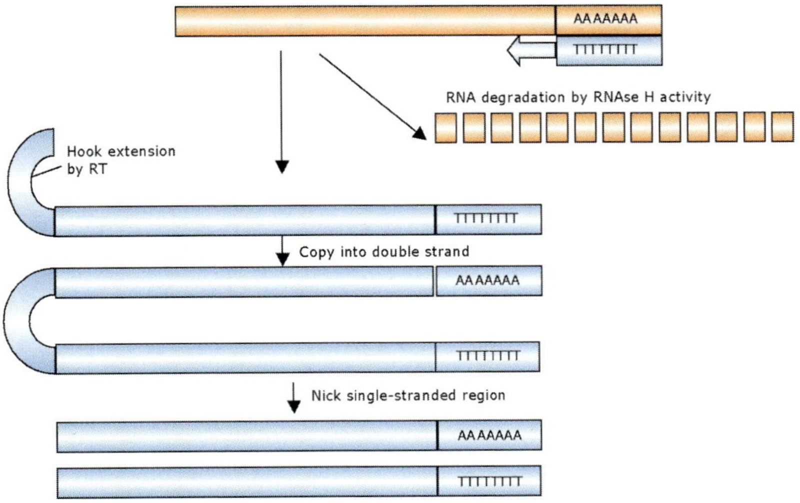

RT-PCR primed by Random hexamers

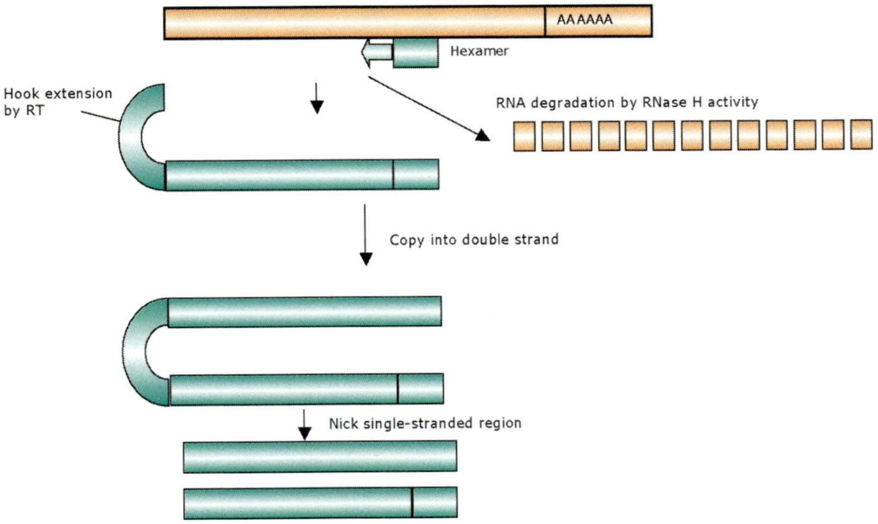

to reduce the artifacts. The **indirect in situ PCR** method, on the other hand, provides additional specificity via use of a probe complementary to an internal sequence within the amplified product, not unlike the one used in real-time quantitative PCR. A variety of probes (double-stranded DNA, single-stranded DNA, 20–30-bp-long oligonucleotides, or single-stranded RNA) have been successfully used. Hybridization of probes to the amplified product follows the same general rules as for in situ hybridization. In in situ PCR, the target is readily accessible because hybridization occurs after DNA amplification.

Suggested Reading

Thaker, V. 1999. In situ RT-PCR and hybridization techniques. Methods Mol. Biol. *115*:379-402.

Cale, J.M., C.E. Shaw, and I.M. Bird. 1998. Optimization of a reverse transcription-polymerase chain reaction (RT-PCR) mass assay for low-abundance mRNA. Methods Mol. Biol. *105*:351-371.

RIBOZYMES

Linked Topics:

Antisense oligonucleotides

Gene targeting

Human Genome Project

Nucleic acid hybridization

Transgenic animals

Various naturally occurring catalytic RNA molecules with an ability to bind to specific mRNAs, including group I introns, are involved in mRNA processing and translation; they bring about sequence-dependent endonucleolytic cleavage and are called ribozymes. Thus, for example, bacterial P RNA is a ribozyme involved in the maturation of the 5′ end of tRNAs. Such naturally occurring catalytic RNAs have been dissected and used as the basis for the rational design of new ribonucleolytic activities. Small ribozymes, among which the best known are the hammerhead and hairpin ribozymes, are potentially tools for genetic manipulation. The hairpin ribozyme–substrate complex contains two independently folding domains that interact with one another to form a catalytic complex, which reversibly cleaves phosphodiesters of RNA substrates to produce fragments with 5′ hydroxyl and 2′,3′-cyclic phosphate termini. While the ligation is 10-fold faster than the cleavage for hairpin ribozymes, cleavage is favored more than 100-fold relative to ligation for the hammerhead ribozyme under the same conditions.

Two different methods of ribozyme delivery are currently in use: exogenous delivery, which involves inserting preformed ribozymes into cells, and endogenous delivery, which requires stable or transient transfection of a ribozyme-encoding gene that would transcribe the ribozyme in the cell itself. The main limiting factor of exogenous delivery for in vivo applications is the instability of ribozymes to nucleases present in the cell or in the extracellular components of blood, skin, and other tissues. The same rules that control intracellular RNA stability and nucleic acid hybridization govern ribozyme function. Chemical modifications like the 2′-hydroxyl groups of the nucleotides by alkylation, substitution with an amino group or a fluorine atom, or changing the phosphate to phosphorothioate internucleotidic linkages, however, allow an increased half-life of ribozymes in serum from minutes to days. As with antisense oligonucleotides, the cellular uptake needs to be improved by the use of cationic liposomes. Endogenous delivery requires efficient delivery, generally via a viral vector, efficient transcription and stabilization, correct folding, and co-localization of the gene product with the target RNA in the cell.

From an intriguing tool to satisfying scientific curiosity, ribozymes have progressed to therapeutic agents with the potential to treat diseases, including the targeting of viral RNAs in infectious diseases, dominant oncogenes in cancers, and specific somatic mutations in genetic disorders. Activated oncogene transcripts, such as the bcr/abl fusion mRNA and ras, are also being targeted with ribozymes. They have also been used for transgenic animal research and gene targeting. Improvements in the nuclease resistance of ribozymes, their catalytic efficiency, and the reduction in their size have all allowed for therapeutic applications for ribozymes. Clinical trials have been initiated using ribozymes targeted against human immunodeficiency virus (HIV). Antiviral efficacy against HIV type 1 has been seen in vitro for both hammerhead and hairpin ribozymes. The Human Genome Project is progressing at a rapid pace. Ribozymes with site-specific cleaving activity could help appreciably in determining the function of open reading frames.

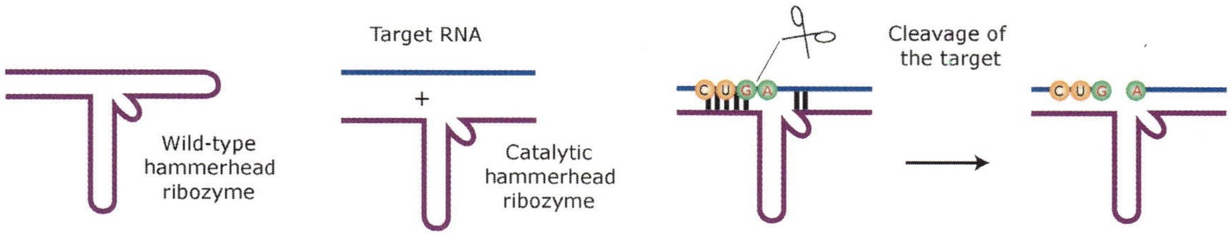

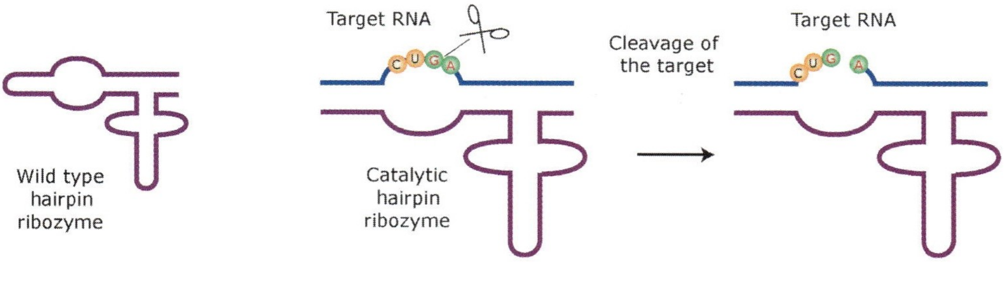

Recently, *trans*-splicing ribozymes containing sequences derived from a group I intron have been employed to refurbish mutated mRNAs in vivo. They can accurately splice in a novel 3′ exonic sequence into a chosen site within a target mRNA molecule. In-frame fusion of the exon results in expression of a new gene product. In principle, the *trans*-splicing ribozymes can be tailored for almost any target mRNA and may provide a new molecular gadget for triggering expression of novel therapeutic genes.

Suggested Reading

Kashani-Sabet, M. 2002. Ribozyme therapeutics. J. Invest. Dermatol. Symp. Proc. *7*:76-78.

Takagi, Y., E. Suyama, H. Kawasaki, M. Miyagishi, and K. Taira. 2002. Mechanism of action of hammerhead ribozymes and their applications in vivo: rapid identification of functional genes in the post-genome era by novel hybrid ribozyme libraries. Biochem. Soc. Trans. *30*:1145-1149.

Blount, K.F. and O.C. Uhlenbeck. 2002. The hammerhead ribozyme. Biochem. Soc. Trans. *30*:1119-1122.

Fedor, M.J. 2002. The catalytic mechanism of the hairpin ribozyme. Biochem. Soc. Trans. *30*:1109-1115.

Ferre-D'amare, A.R. and P.B. Rupert. 2002. The hairpin ribozyme: from crystal structure to function. Biochem. Soc. Trans. *30*:1105-1109.

Freelove, AC. and R. Zheng. 2002. The power of ribozyme technologies: the logical way ahead for molecular medicine and gene therapy Curr. Opin. Mol. Ther. *4*:419-422.

Phylactou, L.A. 2001. Ribozyme gene therapy. Methods Mol. Biol. *175*:521-529.

Doherty, E.A. and J.A. Doudna. 2001. Ribozyme structures and mechanisms. Ann. Rev. Biophys. Biomol. Struct. *30*:457-475.

Kurreck, J. 2003. Antisense technologies. Improvement through novel chemical modifications. Eur. J. Biochem. *270*:1628-1644.

SEQUENCE-TAGGED SITES AND CONTENT MAPPING

Linked Topics:
Alu repeat sequences
Cell hybrids
DNA sequencing
PCR and PCR optimization
Pharmacogenetics and Pharmacogenomics
Plasmids
Single nucleotide polymorphisms

Sequence-tagged sites (STSs) are short (100–400 bp) double-stranded DNA sequences that can be amplified using a specific pair of primers. The starting point for the generation of STS sequences is enriched single chromosome preparations, which are obtained either by flow sorting the chromosomes or from monochromosomal cell hybrids. These chromosomes are digested with restriction endonucleases, and the subchromosomal sequences less than 1000 bp in length are cloned into plasmids. After exclusion of inserts smaller than 100 bp and those with repetitive sequences, random inserts are subjected to DNA sequencing. Suitable primers are then designed for polymerase chain reaction (PCR) amplification of DNA sequences of varying lengths (100–400 bp). Even for the highly complex human genome, the probability of both STS primers annealing to nonspecific locations in the correct orientation is very low, hence a single strong ethidium bromide-stainable band is observed as an STS amplicon. Chromosomal specificity of an STS is confirmed by using an appropriate monochromosomal hybrid, while region-specificity within a chromosome is ascertained by testing against a deletion cell hybrid panel. This region or site on the original genomic DNA is, in a manner of speaking, "tagged" by the ability to assay for a known sequence, hence the name STS.

The STS, thus amplifiable through PCR from a unique location in the human DNA, defines a landmark at a particular subchromosomal genomic site. These STS sequences, therefore, have proven to be a nonvariant currency for physical mapping. Such sequences are deposited in public electronic databases (db), such as **dbSTS**; any researcher wishing to focus on that genomic region can use these STS markers by simply synthesizing the appropriate primers. The STSs can be used readily to construct a set of overlapping clones from any region of interest, provided that the STS are sufficiently closely spaced. The STS primers can be used to screen a library of yeast artificial chromosome (YAC) inserts, where each STS may be amplified in more than one YAC insert (scored as positive), while it may fail to be amplified from other clones (scored as negative). Based on the positive and negative scores, the STS markers can be ordered to obtain an **STS-content map** for the set of overlapping YACs, which then helps in the generation of a **contig**. The problem with the presence of chimeric clones in the YAC libraries, which resulted in delay of STS-content mapping goal for at least 3 years, has been solved by the development of **terminal STS**. These are derived by the **Alu-vector PCR method**, which takes advantage of the frequent presence of Alu repeat sequences in human genomic DNA. In this method, 5′ terminal STS is obtained by amplifying the 100–400-bp insert region flanked by a vector-specific sequence as an upstream primer and an Alu consensus sequence as the downstream primer. Conversely, the 3′ terminal STS is obtained by amplifying a similar 3′ insert region flanked by Alu consensus sequence as an upstream primer and the vector-specific sequence as the downstream primer. Having thus anchored the ends of a clone, the rest of the intervening insert region is used to order the overlapping STS markers.

As of April 2003, a total of 112,526 STSs were mapped in the human genome. Besides serving as landmarks for physical mapping, STSs are also a useful resource for analysis of single nucleotide polymorphisms (**SNPs**), which are fast becoming a tool in pharmacogenetics and pharmacogenomics for the molecular diagnosis of diseases.

Suggested Reading

Perlin, M.W. 1997. Rapid construction of genome maps. Nat. Med. *3*:1046-1048.

Schuler, G.D., M.S. Boguski, E.A. Stewart, L.D. Stein, G. Gyapay, K. Rice, R.E. White, P. Rodriguez-Tome, et al. 1996. A gene map of the human genome. Science *274*:540-546.

Williams, B.D. 1995. Genetic mapping with polymorphic sequence-tagged sites. Methods Cell Biol. *48*:81-96.

SINGLE NUCLEOTIDE POLYMORPHISMS

Linked Topics:

DNA sequencing

Gel-based SNP detection technologies

High-throughput SNP detection technologies

Human Genome Project

Microarrays

Oligonucleotide synthesis

Pharmacogenetics and pharmacogenomics

Single nucleotide polymorphisms (**SNPs**, pronounced "snips") are the most common genetic variations, known to occur once every 100–300 bases in human (and other) genomes. Association of nucleotide sequence variation with heritable phenotypes has been known for years. Identification and cataloguing such SNPs will accelerate the discovery of such associations and pinpoint any population differences. Thus, compared to the traditional approach of pedigree analysis tracking transmission of a disease through a family by following every member over several generations, SNP identification by obtaining DNA samples from a random set of individuals in a population will be much easier. Because <5% of an individual's DNA sequence encodes proteins, most SNPs are found in the noncoding regions. Obviously, SNPs located within a coding sequence are more likely to alter the function of a protein and, hence, would be of special interest to researchers. Nevertheless, those situated around the coding regions will also have tremendous value for charting genetic associations with disease states. Upon discovery, SNPs can help medical research in a number of ways. For example, pharmacogenetics is the field of research that aims to analyze the relationship between genetic variations, such as SNPs that affect drug metabolism, to eventually predict drug response at the level of the individual patient. Large-scale SNP detection is, therefore, one of the important tools of pharmacogenetics.

In collaboration with the National Human Genome Research Institute, the National Center for Biotechnology Information has established the **dbSNP** database, which serves as a central repository for both single base nucleotide substitutions and short deletion and insertion polymorphisms. Once validated, these polymorphisms could be used by additional laboratories, because the data in dbSNP is made freely available to the scientific community via file transfer protocol (FTP) from the SNP homepage (http://www.ncbi.nlm.nih.gov/SNP). As of June 4, 2003, 4,145,633 had been documented in dbSNP. Numerous commercial organizations are also developing such databases, and the extent of the contribution of their output to the medical research community might come at a cost.

Compared to hunting for random mutations in a gene or genome, mutation detection at known sites is easier thanks to strides in DNA sequencing and oligonucleotide synthesis. While diagnostically relevant SNP detection is of obvious value, detecting random SNPs is also likely to reward the field of pharmacogenomics. While gel-based SNP detection technologies have been the workhorses, a number of automated high-throughput SNP detection technologies are fast becoming available, such as microarray, mass spectrometry, capillary electrophoresis, and high-performance liquid chromatography (HPLC)-based DNA fragment analyses, which are capable of simultaneously analyzing several thousand SNPs. While positional cloning based on SNPs accelerated the identification of human disease genes, recent use of high-density oligonucleotide arrays to detect alleles has allowed genotyping of thousands of bi-allelic SNPs in a single experiment.

Ongoing developments in SNP detection methods have forged novel

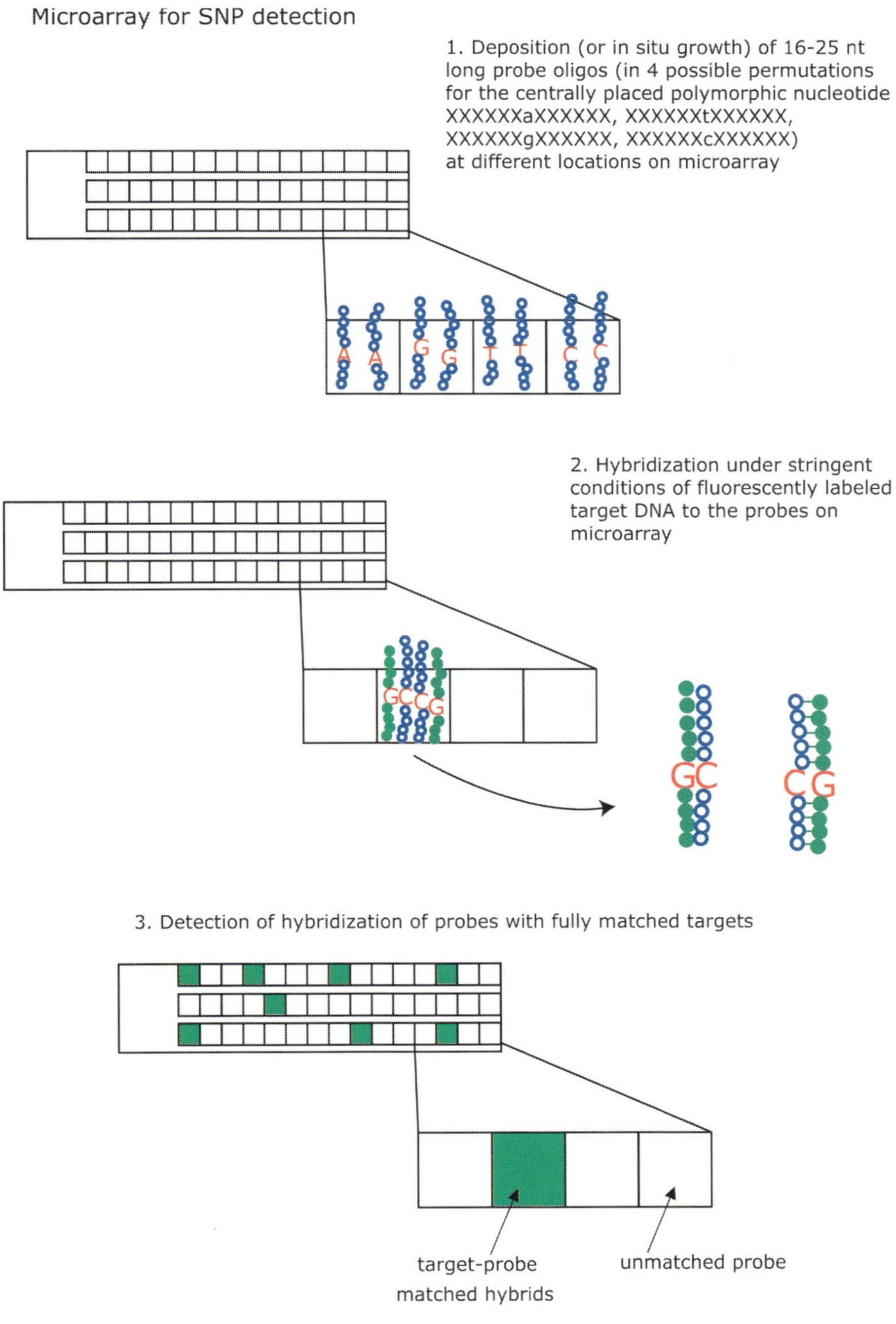

collaborations between biologists, medical professionals, engineers, and bioinformatics specialists, which will have a profound impact on the medical profession and biotechnology industries. The nearly complete sequencing aspect of the Human Genome Project, combined with SNP detection, will have far-reaching implications for human healthcare.

Suggested Reading

Salisbury, B.A., M. Pungliya, J.Y. Choi, R. Jiang, X.J. Sun, and J.C. Stephens. 2003. SNP and haplotype variation in the human genome Mutat. Res. *526*:53-61.

Shastry, B.S. 2002. SNP alleles in human disease and evolution. J. Hum. Genet. *47*:561-566.

SOUTHERN AND NORTHERN BLOTTING AND HYBRIDIZATION

Linked Topics:

Autoradiography

Gene cloning methodologies

Genomic DNA and cDNA libraries

In situ hybridization

Modifications of fluorescent in situ hybridization

Nonisotopic labeling and detection

PCR and PCR optimization

Restriction endonucleases

The technique to transfer size-fractionated nucleic acid molecules from a gel to a filter for analysis was developed by Dr. Edward M. Southern in 1975 and has become an indispensable tool for molecular biologists the world over. This method, developed initially for DNA, is referred to as the Southern blot (transfer) technique.

In this procedure, the target DNA, cleaved with one or more restriction endonucleases, is size-fractionated by agarose gel electrophoresis (whereby DNA fragments, negatively charged due to phosphate groups, are sieved through the porous agarose gel driven from the negative electrode towards the positive electrode), denatured by soaking the gel in alkaline solution (NaOH), and transferred (usually overnight) to a nitrocellulose or a nylon membrane. For this, the gel is placed on an absorbent paper, and a nitrocellulose or nylon (charged or uncharged) membrane the exact size of the gel is laid over it. A high stack of absorbent paper sheets is placed on top of the transfer membrane, and the absorbent paper under the gel is kept wet with concentrated salt solution, such as 20× SSC (3 M sodium chloride, 0.3 M sodium citrate), which flows up by capillary action through the gel to the nitrocellulose or the nylon membrane and into the absorbent paper placed above. The DNA extracted from the gel by the upward movement of the high salt solution adheres to the nitrocellulose or the nylon membrane. The location of the individual DNA fragments on membrane is a faithful representation of the size separation achieved by agarose gel electrophoresis. The alternative methods for transfer of DNA from gel onto membrane are downward capillary transfer or employment of vacuum or electrophoretic force (electroblotting). The transferred DNA is immobilized by a brief exposure to UV light.

In its single-stranded membrane-bound form, the target DNA can be analyzed via its hybridization (annealing) with the single-stranded DNA probe. If the probe is labeled with radioisotopes (P^{32} or S^{35}), the hybridization is detected by autoradiography. Alternatively, specific hybridization can also be identified using nonisotopic labeling and detection methods. The Southern blot hybridization technique has proved to be extremely useful in analytical procedures for detection and copy-number determination of specific sequences in the complex genomic DNAs, confirming DNA cloning results or screening of genomic libraries. Major advances in molecular diagnostics, such as restriction fragment-length polymorphism (RFLP) analysis, restriction landmark genomic scanning (RLGS), and DNA typing, also employ Southern transfer as an integral part.

Northern blot hybridization (by an extension of "Southern" nomenclature analogy) is a variant of Southern blotting, in which the target nucleic acid is RNA instead of DNA. In this case, however, the denaturation of RNA species in the gel is brought about by organic denaturants like

Nucleic Acid Hybridization: Southern Blot Hybridization

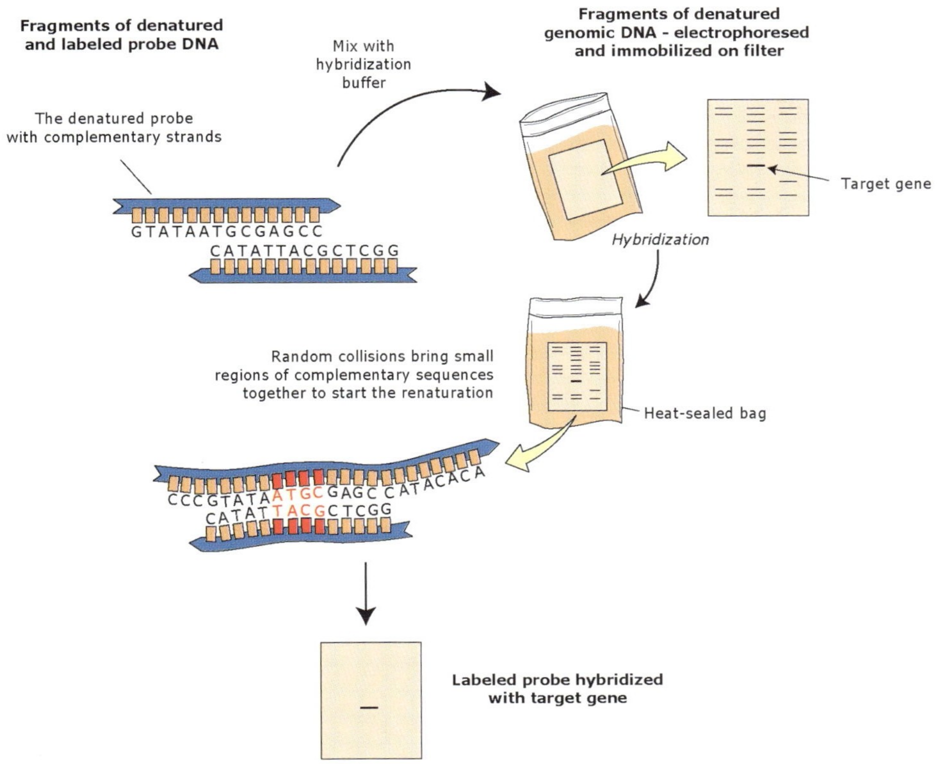

formaldehyde or glyoxal instead of alkali, because the latter hydrolyzes RNA. The principal use of this method is to obtain information on the expression patterns of specific genes. A short stretch of the target gene can be used as a probe and hybridized against a Northern blot containing, in different gel lanes, samples of RNA isolated from a variety of different tissues. The data obtained can provide information regarding expression and the relative abundance of the transcripts for a given gene in a range of cell types. In addition, by revealing transcripts of different sizes, it may provide evidence for the use of alternative regulatory gene sequences, such as promoters, splicing sites, or polyadenylation signals.

An alternative to Southern hybridization for detection of DNA sequences is amplification of the target sequence by polymerase chain reaction (PCR). Similarly, alternatives for Northern hybridization for detection of a given RNA are ribonuclease protection assay and reverse transcription PCR (RT-PCR). While RT-PCR can detect the presence of an RNA molecule, the size the given RNA molecule can be determined by ribonuclease protection assay. The intracellular or chromosomal localization of both specific DNA and RNA sequences can be determined using the techniques of in situ hybridization or fluorescent in situ hybridization (FISH).

Suggested Reading

Kendall, L.V. and L.K. Riley. 2000. Southern and Northern hybridization. Contemporary Top. Lab. Anim. Sci. *39*:50.

SUBTRACTIVE HYBRIDIZATION

Linked Topics:
PCR and PCR optimization
Positional cloning (reverse genetics)

Subtractive hybridization methods are used to isolate genes that are specifically expressed in one cell type or a specific tissue, including genes that are differentially regulated during activation or differentiation of the cells, besides the genes that are involved in the development of diseases such as cancer or viral infections. Several different strategies have been developed to achieve this goal. They are all based on subtracting the common transcripts between different cell types or tissues, leaving behind the unique cell-type or tissue-specific transcripts for further analysis. Subtraction protocols have typically involved hybridization of mRNA from the target cell population to first-strand cDNA from the subtractor cell population. The duplex mRNA–cDNA hybrids are separated from unhybridized unique target mRNAs. While the separation was traditionally achieved by hydroxyappatite column chromatography, magnetic separation is increasingly being used to this end. One way to proceed from the unhybridized unique single-stranded mRNAs is to make first-strand cDNA and employ the cDNA strands in screening of cDNA libraries or convert the first strands into double-stranded cDNA and clone into a suitable vector. Another approach is to make immobilized cDNA libraries from both the target mRNA population and the subtractor mRNA independently. A major requirement of successful subtractive hybridization is a large amount of poly(A)+ mRNA and highly purified single-stranded cDNA. This technique is demanding and labor-intensive. One of the problems with the subtractive hybridization technique is that specific mRNAs or cDNAs from gene families or repetitive sequences are sub-

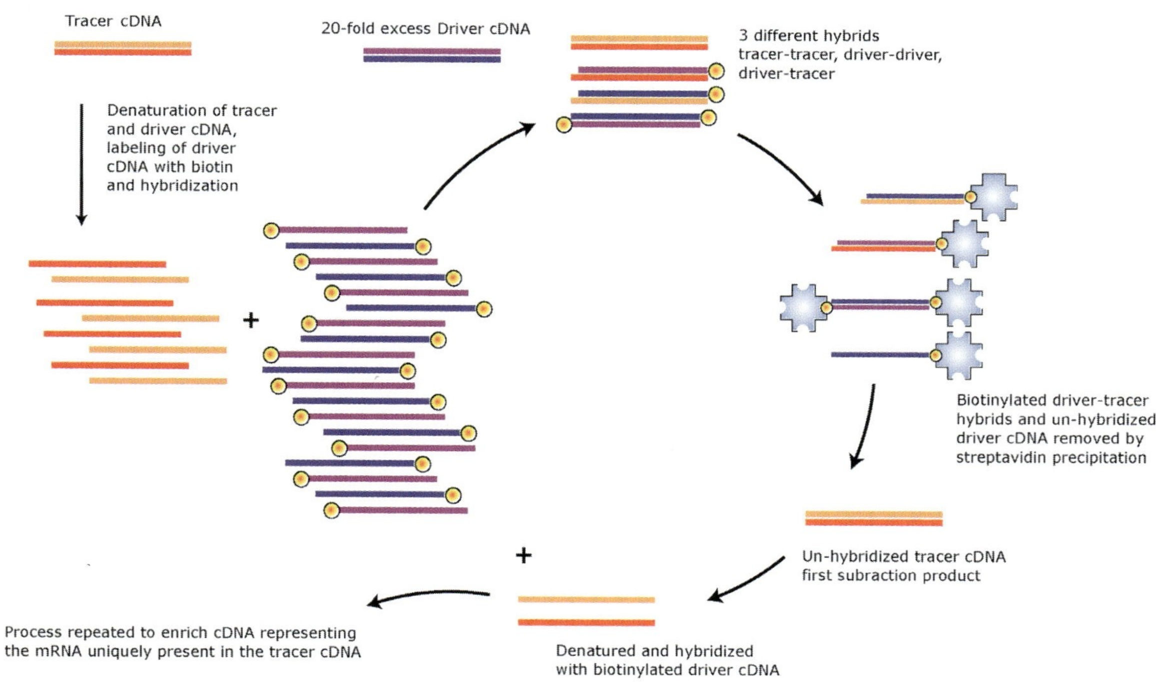

tracted away by common sequences, which can be circumvented to some extent by using short random primed segments from target mRNAs.

However, considering that at any given moment thousands of genes are likely expressed in a mammalian cell, only a small fraction (approximately 10%) accounts for almost 30% of all transcripts. Conversely, a very large number of genes are expressed at extremely low levels. Straightforward subtraction may not reveal genes that are differentially expressed at only a very low level. Polymerase chain reaction (PCR)-based **suppression subtractive hybridization** (SSH) is a technique that combines normalization and subtraction. SSH defines the subtle transcriptional differences between two genomes. PCR amplification is used to enrich for unique segments of restricted DNA, while simultaneously limiting nontarget amplification by suppression PCR. The genome of interest in this comparison is called the tester DNA, and the comparison genome is called driver DNA. The abundant transcripts are suppressed, while rare transcripts are enriched to the same order of magnitude by normalization. The subtraction then allows differentially expressed transcripts to be isolated. This technique is comparable to representational difference analysis (RDA).

Suggested Reading

Byers, R.J., J.A. Hoyland, J. Dixon, and A.J. Freemont. 2000. Subtractive hybridization–genetic takeaways and the search for meaning. Int. J. Exp. Pathol. *81*:391-404.

TAQMAN® AND MOLECULAR BEACONS

Linked Topics:

Fluorescence resonance energy transfer

Oligonucleotide synthesis

Single nucleotide polymorphisms

While unlabeled primers are used to prime the polymerase chain reaction (PCR), an additional fluorophore-quencher-labeled oligonucleotide probe (modified at the 3′ end to prevent mispriming and extension) is used. During PCR, if the target of interest is present, the probe anneals specifically between the forward and reverse primer sites of extended PCR product in the annealing phase. As the next primer extension proceeds through the probe-bound site, the 5′-3′ exonuclease activity of *Taq* DNA polymerase excises the probe nucleotides sequentially from 3′ to 5′, thus releasing the fluorescent nucleotide away from the quencher. A form of fluorescence resonance energy transfer (FRET), this process is commonly referred to as **TaqMan** chemistry. Due to the cyclic cleavage of the reporter molecules, the fluorescence intensity of the overall system increases in direct proportion to the amount of DNA amplified. Built-in fiber optic cables train a laser beam into each of the PCR sample wells or capillaries in the thermal cycler. The resulting fluorescent emission returns via the optical fibers and is directed to a spectrograph with a charged-couple device (CCD) camera, which records it continually, and data are plotted in real time as the function of the cycle number. TaqMan probes can also be designed to detect single nucleotide polymorphisms (SNPs). While one probe is fully complementary to the target, the alternative probe contains a centrally located base pair mismatch, sufficient to destabilize the

TaqMan

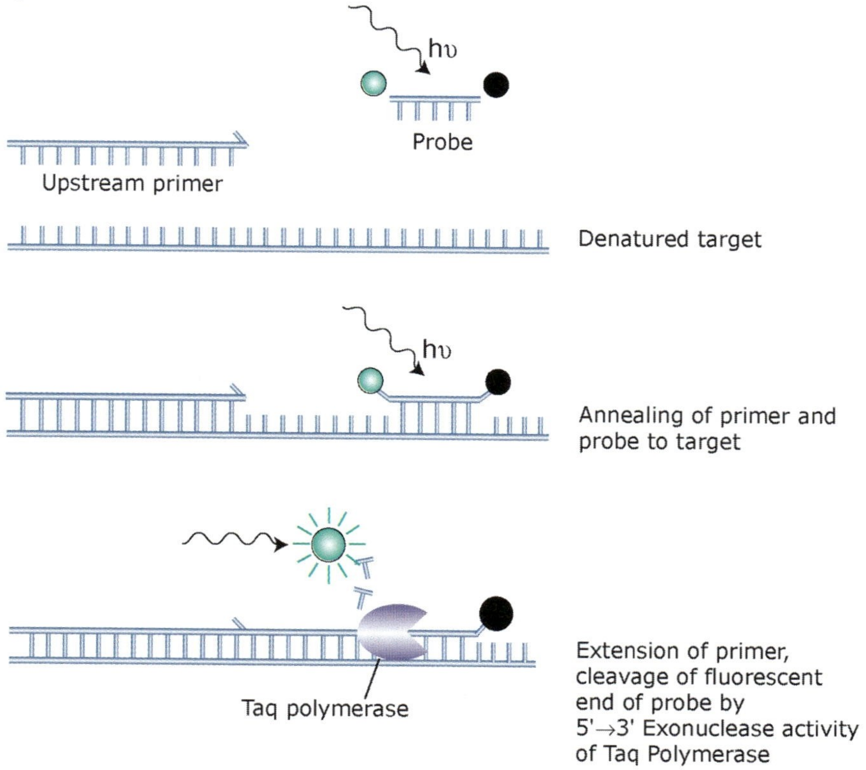

20-bp oligonucleotide probe. In the absence of a SNP, the complementary probe anneals to its target and is cleaved to produce a real-time PCR signal. The probe with a mismatch, however, either fails to bind to the target or falls off the target DNA strand before cleavage of the reporter dye. Since the fluorescent moiety never separates from the quencher moiety, a signal is not observed with the mismatch probe. Alternatively, since the fluorescence signal can also be generated if FRET occurs between two adjacent fluorophores, two hybridization probes are designed to anneal one base apart with the adjacent internal target DNA sequences. These are fluorescently labeled, one with a donor fluorophore at the 3′ end and another with an acceptor fluorophore at the 5′ end. The resulting FRET due to proximity of the fluors is monitored by a micro-volume fluorometer integrated with a thermal cycler. In addition to detecting SNPs, either of these methods can also be used to identify small gene deletions and insertions.

With **molecular beacons**, each primer has a fluorophore and a quencher at or near the ends. They are designed such that each forms an intramolecular hairpin-like secondary (stem loop) structure which, when unannealed from the target DNA, cannot fluoresce due to FRET. Formation of a stable duplex between molecular beacons and templates during every primer extension results in a dramatic increase in fluorescence due to separation of the fluorescent dye from the quencher. Mismatched probe-target hybrids dissociate at substantially lower temperatures than perfectly matched hybrids. This thermal instability of mismatched hybrids adds to the specificity of molecular bea-

Molecular Beacons

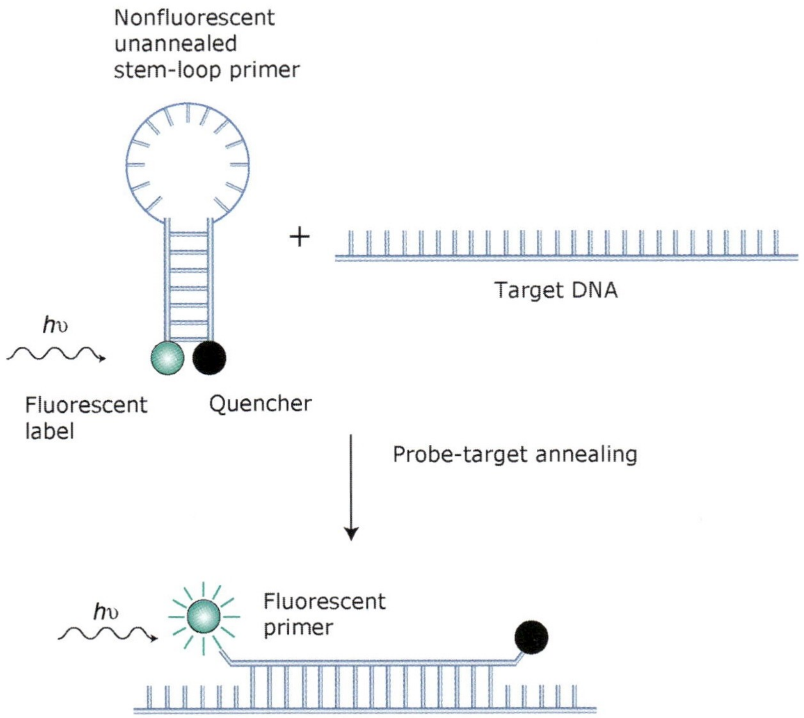

cons. During PCR, formation of a longer and more stable primer–template hybrid is favored to the initial stem-loop structure, precluding the simultaneous existence of both structures. As the PCR proceeds, increasing participation by the primers in formation of products increases the fluorescence, which is continuously monitored by fluorescence detection sensors. The recently developed "**uniprimer**" consists of a 3′ 18-base oligonucleotide "tail" and a 5′ stem-loop sequence labeled with the fluorochrome–quencher pair. The tail sequence is also incorporated at the 5′ end of one of the target-specific primers, ensuring participation of the uniprimer in subsequent cycles, which thus can act as a universal PCR primer. For SNP detection, two molecular beacons with exact sequence matches to the wild-type and variant alleles are used in the same PCR. These two molecular beacons, labeled with different fluorophores that have distinct emission wavelengths, allow simultaneous determination of three possible allelic combinations. Recent developments in DNA synthesis allow introduction of both fluorescent moiety at the 5′ end and quencher moiety at the 3′ end during production, making it possible to synthesize a molecular beacon completely on a DNA synthesizer.

Suggested Reading

Antony, T. and V. Subramaniam. 2001. Molecular beacons: nucleic acid hybridization and emerging applications. J. Biomol. Struct. Dyn. *19*:497-504.

Fang, X., Y. Mi, J.J. Li, T. Beck, S. Schuster, and W. Tan. 2002. Molecular beacons: fluorogenic probes for living cell study. Cell Biochem. Biophys. *37*:71-81.

TELOMERES AND TELOMERASE

Linked Topics:
Antisense oligonucleotides
Aptamers
Ribozymes

Each time a cell divides, it loses 50 to 100 bp at the ends of its DNA, due to an inherent flaw in the way cells copy their genomic material. Described first by James Watson, the end replication problem is a consequence of polarity of DNA strands and the resulting lagging strand formation in the process of DNA synthesis. In human cells, telomeres are 5000–15,000-bp-long sequences at chromosomal ends, composed of repeats of [TTAGGG]n. About 50–100 of these are lost with each cell division. Although the number of repeats of telomeres vary from species to species, their functions are universal and include protection of chromosomes from DNA degradation, maintenance of nuclear structure, prevention of end-to-end fusions, and rearrangements. The telomeric sequence has a 3′ G-rich single-stranded overhang that tucks back into the double-stranded part of the telomeric sequence, protecting the end of the chromosome from recombination events with other chromosomes and from exonuclease digestion. Telomeres are also considered to be anchoring points of chromosomes to the nuclear envelope.

During the past several years, there has been substantial evidence that cellular aging depends on the number of cell cycles and resulting loss of telomeric repeats. Owing to the progressive loss of telomeric sequences in successive cell division cycles, telomeres shorten with each division, causing genetic imbalance. Evidence from many studies has established that the shortening of telomeres whereby they reach a critical length, acts as a mitotic clock, causing growth arrest and eventual cell death; this process probably forms the mechanistic basis of the "Hayflick limit" of cell divisions in somatic cells, named after the discoverer, Leonard Hayflick. Most normal somatic cells derived from adults are limited in the number of times they can divide. The number of replicative events that a cell or cell line can undergo before reaching replicative arrest [the state where cell division can no longer occur without additional mutation(s), also arguably called replicative senescence] is known as the Hayflick limit. However, given some recent data, the exclusive role of telomere shortening in induction of cell growth arrest is controversial. It is possible that chronic mild oxidative stress may also be an important player in this growth arrest, also known as stasis. In fact, such an oxidative stress accelerates telomere shortening, suggesting that, on top of the end-replication problem, oxidative stress is a major cause of telomere shortening. Cells circumvent the progressive loss of telomeres through several strategies. The most common method is adding telomeres back to the chromosomal ends by using an enzyme complex termed telomerase.

Telomerase is a ribonucleoprotein enzyme complex composed of an RNA [human telomerase RNA (hTR)] component complementary to a hexameric telomeric sequence TTAGGG and a reverse transcriptase enzyme component (human telomerase reverse transcriptase (hTERT). Telomerase ribonucleoprotein binds to the 3′ overhang of an existing telomere sequence such that the overhang base pairs with the template hTR RNA. The hTERT activity then extends the telomere sequence copying from the template region of hTR. Once the copying is completed for a unit length of template, the telomerase translocates to realign the new 3′ end to the start of the templating region, eventually adding multiple hexamer repeat sequences. During this process, telomerase acts in concert with telomere binding proteins. Telomerase is not easily detected in most somatic tissues of adult vertebrates. Telomerase activity, which is highest in germline cells, progressively tapers off as cells differentiate. Of enormous clinical interest and the focus of considerable study in the past few years is the reactivation of telomerase in the vast majority of human cancers, including cancers of breast, lung, prostate, stomach, pancreas, thyroid, colon, cervix, and bladder. Telomerase reactivation probably plays a permissive role for acquisition of all the mutations necessary to acquire and/or retain the cancerous phenotype. It is for these characteristics that telomerase has emerged as a potential marker for detection and prognostication, and a target for adjunct cancer therapy.

The current options for functional inhibition of telomerase include use of antisense therapeutics such as antisense oligonucleotides, aptamers, or ribozymes directed against hTR, and synthetic peptides or antibodies directed against the protein component hTERT.

Suggested Reading

Odago, F.O. and S.L. Gerson. 2003. Telomerase inhibition and telomere erosion: a two-pronged strategy in cancer therapy. Trends Pharmacol. Sci. *24*:328-331.

Incles, C.M., C.M. Schultes, and S. Neidle. 2003. Telomerase inhibitors in cancer therapy: current status and future directions. Curr. Opin. Invest. Drugs *4*:675-685.

Zglinicki, T., J. Petrie, and T.B.L. Kirkwood. 2003. Telomere-driven replicative senescence is a stress response. *21*:229-230.

TOUCH-DOWN PCR

Linked Topics:

Modifications of PCR

PCR and PCR optimization

Touch-down polymerase chain reaction (PCR) involves decreasing the annealing temperature by 1°C every cycle or every second cycle, so as to touch down to the annealing temperature. The reaction is then continued at the annealing temperature for 10 or more cycles. This technique was originally intended to bypass the more complicated optimization process for determining optimal annealing temperatures, even if the degree of primer–template complementarity is not fully known. The principle behind this method is that any differences in the thermal melting temperature (T_m) between correct and incorrect annealing gives a twofold difference in product amount per cycle. Thus, one achieves enrichment of the correct product over any incorrect products in a one-step procedure without optimization. It has been demonstrated that priming initiates even above the optimum annealing temperature, which helps to ensure a competitive advantage for the target amplicon. On the other hand, annealing temperatures well below the T_m can also significantly increase yields without promoting spurious amplification. A modification of touch-down PCR, termed step-down PCR, consists of fewer but steeper incremental declines in annealing temperature and can simplify thermal cycler programming. Both of these variations serve to minimize the need to optimize annealing temperature or buffer conditions and yet produce strong bands of amplification products in the gel.

The utility of this technique in molecular diagnostics has been shown for genetic diseases such as spinocerebellar ataxia (SCA). SCA is characterized by unstable expansion of a CAG repeat in the *SCA2* gene, coding for a polyglutamine stretch in the corresponding protein. Touch-down PCR results in amplification of both normal and pathological alleles simultaneously, which can be detected by denaturing polyacrylamide gel electrophoresis. Thus, this method has a potential diagnostic role even in other diseases, which are characterized by anomalous nucleotide repeat expansions, such as Huntington's disease and muscular dystrophy, and even in cancers showing microsatellite instability.

Another use for this procedure is in determining DNA sequence for a known peptide sequence. The strategy uses two sets of degenerate primers matching potential coding sequences at the two ends of a peptide of known length. A touch-down PCR using these degenerate primers results in a large number of products; but since the exact interprimer distance is known from the peptide sequence, it is possible to select the desired product based on size, and subsequently clone and sequence the same. Yet another application of touch-down PCR is in identifying and isolating a homologous DNA sequence from distant genomes. Degenerate PCR primers can be synthesized based on known peptide sequences of a given protein from one organism, and using these in touch-down PCR with genomic DNA from another organism as template, related sequences can be identified and enriched.

Suggested Reading

http://bip.weizmann.ac.il/mb/bioguide/pcr/PCRTouchdown.html

TRANS-ACTING ELEMENTS: TRANSCRIPTIONAL FACTORS

Linked Topics:

Cis-*regulatory elements*

DNA binding proteins

Homeobox genes

Two-hybrid system

Trans-acting elements (relative to *cis*-regulatory elements within a gene) are transcription factor (TF) proteins that recognize and bind short nucleotide sequences in a gene. The binding usually results from extensive correspondence, somewhat like lock-and-key, between the surface of a domain of the protein and the surface features of the DNA double helix in a specific region. They are thought to home in on the target sequence after repeated collision with DNA until they stumble upon a functional site within a promoter. Although individual interactions between amino acids and nucleotides are mostly weak hydrogen bonds or ionic bonds, the region of DNA–protein binding is typically characterized by about 20 such contact points, which collectively ensure a strong and specific binding. Many conserved structural motifs have been identified in eukaryotic TF with varied specificity. These motifs mostly use α helices, and in some cases, such as eukaryotic p53 β sheets, to bind to the major groove of the DNA. There are also examples of TF that bind to the minor groove, like OCT-1 POU, which bind homeobox genes, and high mobility group (HMG) proteins that bind T cell receptor (TCR)-α enhancer. Many TF fine tune their regulatory potential by homodimer versus heterodimer formation. In the latter, each monomer binds to a half-target sequence, allowing a combination of unique controls. An example is eukaryotic JUN protein, which binds weakly to the AP-1 promoter as a jun-jun homodimer, but tightly as a jun-fos heterodimer. The oligomeric nature of a TF can be examined by techniques like DNase I footprinting and electrophoretic mobility shift assay (EMSA). Another technique to identify protein–protein association is the **two-hybrid system**.

Typically, two distinct functional domains can be identified in a transcription factor: **DNA-binding domain** and **activation domain**. Specific association between DNA-binding domains and activation domains forms the basis of the two-hybrid system. There are four common structural motifs seen in transcriptional factors.

Leucine zipper motif is a helical peptide stretch rich in leucine residues, which usually occurs once every seven amino acids (once every two turns of the helix). This readily forms a dimer, where each monomer consists of an amphipathic α helix, such that hydrophobic side groups of the constituent amino acids face one way, while polar groups face the other. The overall dimer is a Y-shaped structure that grips the double helix. Besides forming homodimers, leucine zipper proteins can also form heterodimers. Such heterodimer formations provide an important combinatorial control mechanism in gene regulation.

Helix-loop-helix motif consists of two α helices, one short and one long, connected by a flexible loop. The loop permits folding back, such that the two helices can stack against each other in parallel planes.

Helix-turn-helix motif is a common motif found in homeobox genes and many TF. It consists of two short α helices separated by a short

DNA Footprinting

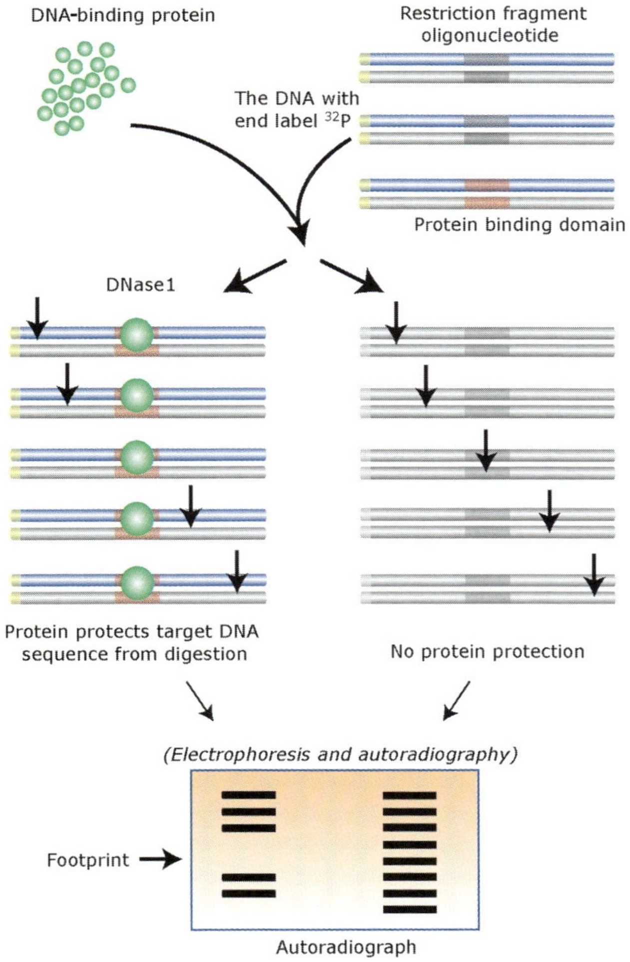

amino acid sequence that induces flexure, such that the two α helices lie in different planes.

The zinc finger motif involves the binding of a zinc ion by four conserved amino acids (four cysteine residues or two cysteine and two histidine residues) so as to form a loop. The primary contact is made by an α helix binding to the major groove. This cys2/his2 finger typically comprises about 23 amino acids, with neighboring fingers separated by a stretch of about seven or eight amino acids.

Suggested Reading

Ladomery, M. and G. Dellaire. 2002. Multifunctional zinc finger proteins in development and disease. Ann. Hum. Genet. *66*:331-342.

Jamieson, A.C., J.C. Miller, and C.O. Pabo. 2003. Drug discovery with engineered zinc-finger proteins. Nat. Rev. Drug Discov. *2*:361-368.

Luscher, B. and L.G. Larsson. 1999. The basic region/helix-loop-helix/leucine zipper domain of Myc proto-oncoproteins: function and regulation. Oncogene *18*:2955-2966.

Urnov, F.D. 2002. A feel for the template: zinc finger protein transcription factors and chromatin. Biochem. Cell Biol. *80*:321-333.

TRANSCRIPTION-MEDIATED AMPLIFICATION

Linked Topics:
Modifications of PCR

Transcription-mediated amplification (TMA), also variously known as transcription-based amplification system (**TAS**), self-sustained sequence replication (**3SR**), or nucleic acid sequence-based amplification (**NASBA**), is the technology that sequentially uses RNA polymerase and reverse transcriptase to autocatalytically drive the nucleic acid amplification reaction. TMA, which can amplify either DNA or RNA, produces RNA amplification products and is an isothermal alternative to polymerase chain reaction (PCR), such that the reaction is performed at the same temperature in its entirety.

To start with, the first primer [a **promoter-primer**, which is an oligo(dT) sequence or a gene-specific sequence synthesized to be collinear with a T7 promoter sequence 5′-AATACGACTCACTATAGGGAGA-3′] anneals to the mRNA target and is extended by reverse transcription (RT). The RNA strand is degraded by RNase H activity. A second upstream primer then anneals to the DNA strand and is extended by RT, yielding a DNA template with an active double-stranded T7 promoter sequence at its end. T7 RNA polymerase binds to this promoter and transcribes the template strand, yielding 100–1000 RNA copies. The upstream primer anneals to each of these resultant RNA copies and is extended by RT. Once again, the RNA strand is degraded by RNase H activity, the promoter-primer binds to the first DNA strand, and is extended to yield an active T7 promoter. More RNA copies are made yet again, as T7 polymerase transcribes the double-stranded DNA, and the cycle continues in an autocatalytic fashion. TMA has very rapid kinetics, resulting in up to a 10-billion-fold amplification within an hour or less. Combined with the hybridization protection assay (**HPA**, see below), TMA can permit single closed-tube detection of the target molecules devoid of wash steps. The single tube format also minimizes possibility of contamination and simplifies the procedure.

Hybridization protection assay: A stable fully hybridizable acridinium ester label is employed in HPA, which permits a chemiluminescence-based distinction between hybridized and unhybridized probes. The tricyclic aromatic ester label is attached to the probe via an N-hydoxysuccinimide link. In the presence of a peroxide anion, the label on hybrids is cleaved at position 9 of the acridinium ring. Under alkaline conditions, following formation of an intermediate, excited N-methylacridone is formed, which emits light upon relaxation to the ground state. The acridinium ester label has a limit of detection of about 5×10^{-19} mol (0.5 amol). The labeled probe has an innate characteristic, which allows discrimination between hybridized and unhybridized probes. The acridinium ester moiety intercalates into and protects double-stranded hybrids (hence the name HPA). When the labeled probe is reacted with sodium tetraborate, unhybridized ester molecules are hydrolyzed. Therefore, only the hybridized probes emit light, “flagging” the presence of target sequences. The photon emission as a result of

Transcription-Mediated Amplification (TMA)

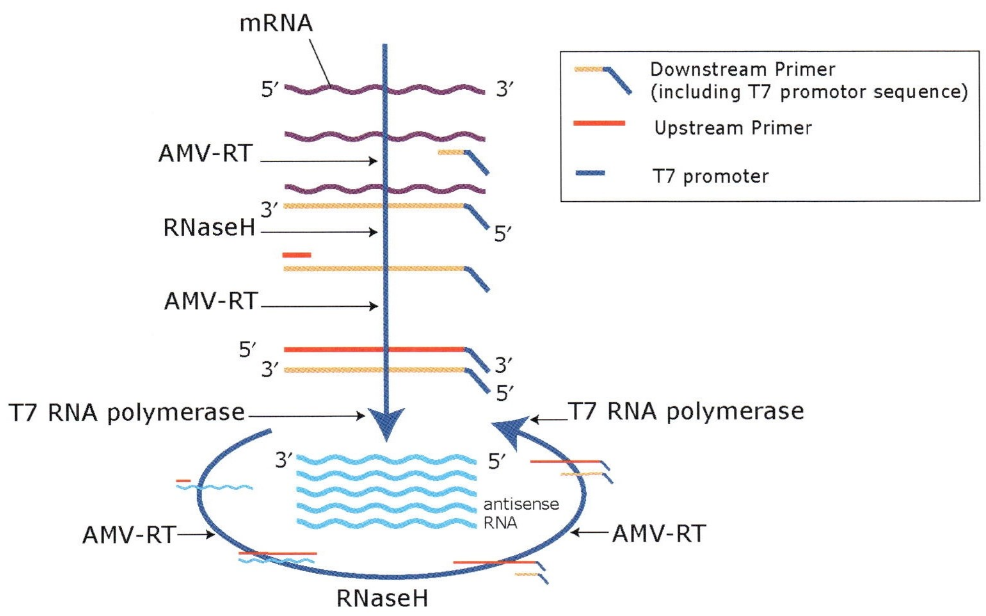

chemiluminescence is measured with a luminometer that injects hydrogen peroxide and alkaline solution.

The isothermal single-tube TMA assay can be very convenient in the clinical laboratory setting. It can be used qualitatively for determination of the presence or absence of a target, or quantitatively to quantify the amount of starting target over a concentration range of several orders of magnitude. It can be used to amplify a single target or put together a multiplex amplification assay to simultaneously detect multiple targets. Applicability of TMA is demonstrated in the detection of infectious diseases, hematologic diseases, and human leukocyte antigen (HLA) typing.

Suggested Reading

Benjamin, R.J. 2001. Nucleic acid testing: update and applications. Semin. Hematol. *38*(Suppl. 9):11-16.

Nelson, N.C. 1998. Rapid detection of genetic mutations using the chemiluminescent hybridization protection assay (HPA): overview and comparison with other methods. Crit. Rev. Clin. Lab. Sci. *35*:369-414.

TWO-HYBRID OR INTERACTION TRAP SYSTEM AND ITS MODIFICATIONS

Linked Topics:

DNA-binding proteins

Trans-*acting elements*

Identification of ligand-receptor interactions is pivotal to pharmacological intervention of human diseases in medicine. Two-hybrid technology is a clever way to accomplish the discovery of such interactions and, thereby, candidate genes for different diseases. It is based on a peculiar feature of modularity in certain transcription factors (like the yeast factor GAL4). These factors have two distinct domains in their structure: (*i*) the DNA binding domain (generally denoted as D) is used for binding with specific DNA sequences at the target gene promoter, and (*ii*) the activation domain (referred to as A) for stimulating transcription. These two domains are connected by a peptide chain, but do not otherwise interact with each other. Thus, the two domains can result in activation of transcription by being merely close to each other although they may not be along the same polypeptide chain. The technique of interaction trapping uses this peculiarity to identify any unknown protein that interacts with a known protein. It is possible to make truncated versions of *GAL4* gene, one of which codes for D domain, and the other which codes for the A domain.

A chimeric gene is made encoding a fusion protein that contains the GAL4 domain D coding sequence fused to a known "bait" protein, say Y, or some segment of the protein Y. Another construct is made between coding region for GAL4 domain A and members of a cDNA library, one of which may encode the protein X ("prey"). Both the chimeric plasmids are cloned into a common host. Also included is a reporter gene, which is under the control of GAL4 promoter. Since protein Y–protein X interaction will bring GAL4 D and A domains together, the host cell harboring such co-expression will express the reporter gene. Having identified a protein that is crucially involved in pathogenesis, it is possible to identify the genes for all other proteins that interact with the first protein.

Reverse two-hybrid system, on the other hand, enables genetic selection against a specific protein–protein interaction, permitting identification of molecules or mutations that dissociate known interactions. Thus, yeast strains are engineered in which the interaction of two given proteins co-expressed is deleterious to growth, while the dissociation of the interaction provides a selective growth advantage. One application of this modification is that it may be possible to identify dissociator molecules that could be used as therapeutic agents.

Adapting from the basic two-hybrid system, the three-hybrid system (two-bait system) has evolved where a third synthetic bifunctional hybrid ligand is included. This system can be used to discover receptors for small ligands and to screen for new ligands to known receptors. It can also be used to analyze RNA–protein interactions in which the binding of a bifunctional RNA to each of two-hybrid proteins can activate transcription of a reporter gene in vivo. Since the latter assay relies on the physical properties of the RNA and protein and not on their natural biological activities, it has broad application potential in identification of interactions of RNA-binding proteins and RNAs.

There is, of course, the need to understand the false positive interactions between proteins. Some interactions seen in yeast may not necessarily occur in nature, because although they are expressed in the same cell, the two proteins may not be expressed in the same subcellular compartment or in the same time frame. Alternatively, the prey proteins may show promiscuous interaction to unrelated baits, or they may interact, not with the promoter sequences of reporter genes directly, but via other proteins which do so. Various safeguards are, therefore, routinely included in these assays.

Two-Hybrid System

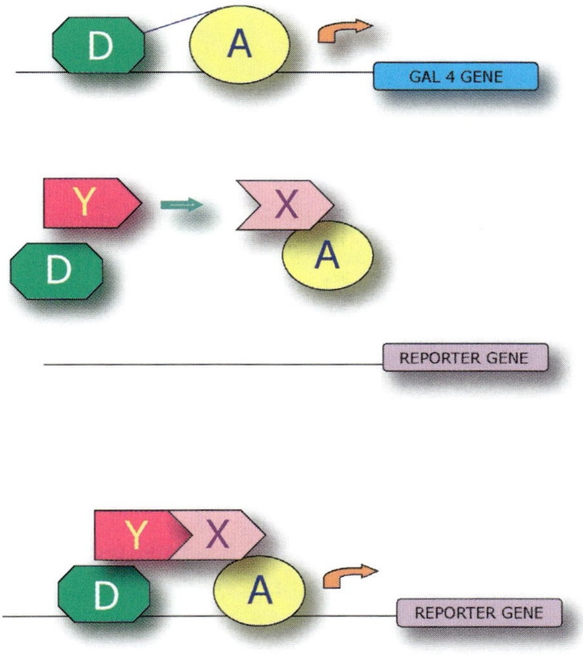

Suggested Reading

Coates, P.J. and P.A. Hall. 2003. The yeast two-hybrid system for identifying protein-protein interactions. J. Pathol. *199*:4-7.

Auerbach, D., S. Thaminy, M.O. Hottiger, and I. Stagljar. 2002. The post-genomic era of interactive proteomics: facts and perspectives. Proteomics *2*:611-623.

Toby, G.G. and E.A. Golemis. 2001. Using the yeast interaction trap and other two-hybrid-based approaches to study protein-protein interactions. Methods *24*:201-217.

YEAST ARTIFICIAL CHROMOSOMES

Linked Topics:

Bacterial artificial chromosomes

Bacteriophages

Human Genome Project

P1-derived artificial chromosomes

Physical mapping

Plasmids

While DNA sequences ranging in size from 50–350 kb can be cloned in vectors such as bacteriophages, bacterial artificial chromosomes (BACs), P1-derived artificial chromosomes (PACs), or P1 vectors, the most popular system for cloning very large (megabase-sized) DNA fragments is the construction of yeast artificial chromosomes (YACs). In principle, cloning in yeast cells offers some advantages over the cloning in bacterial cells. Certain eukaryotic sequences, particularly ones that include repeat sequences, are difficult to propagate in bacterial cells that do not have such DNA elements. But, being eukaryotic, yeast cells tolerate such sequences better. In addition, YACs are able to clone very large DNA fragments. This feature was developed after the realization that a substantial portion of yeast chromosomal DNA is not required for normal function. The length of yeast genomic DNA necessary for functional activity in vivo is only a few hundred base pairs of DNA.

The essential functional components of yeast chromosomes are the centromere, telomeres, and autonomous replicating sequences (ARSs). The centromere elements (CEN) are required for disjunction of sister chromatids in mitosis and of homologous chromosomes in first meiotic division. Telomere sequences (TEL) are required for complete replication of linear molecules and the protection of ends of chromosomes from nuclease attack, while the autonomous replication sequences act as specific replication origins. The tRNA transcript of the yeast gene *SUP4* suppresses *ochre* mutation in AB 1380 strain of *Saccharomyces cerevisiae*, which gives white colonies. Insertional inactivation of *SUP4* by cloning a foreign DNA stretch in *Eco*RI site results in nonsuppression of the *ochre* mutation and formation of red yeast colonies.

Hence, a novel cloning system was developed based on the use of chromosomal replicons, or ARS elements, as an alternative to the ubiquitous use of cloning vectors based on extra-chromosomal replicons (such as those found in plasmids and bacteriophages). To make a YAC, it is only necessary to combine the yeast genetic elements comprising of two TEL, one CEN, one ARS, one SUP4, and a selectable growth marker gene element (URA3). These can all be cloned in a plasmid containing ColE1 plasmid replication origin, an ampicillin resistance (amp^r) gene, and grown in bacterial cells until required for recombination.

A suitably sized large foreign DNA fragment can be obtained by pulsed field gel electrophoresis (PFGE) size fractionation, and inserted into *Eco*RI site within *SUP4* to obtain a recombinant linear YAC molecule with terminally positioned telomeric sequences. Since such a construct cannot be engineered in situ directly into yeast cells, yeast cells have to be treated in such a way as to remove the external cell walls. The exogenous constructs are then taken up by the resulting yeast spheroplasts. The spheroplasts have to be embedded in agar, since they are osmotically unstable.

The overall efficiency of transformation is extremely low, the yield of cloned DNA being only one copy per cell. Nevertheless, the capacity to

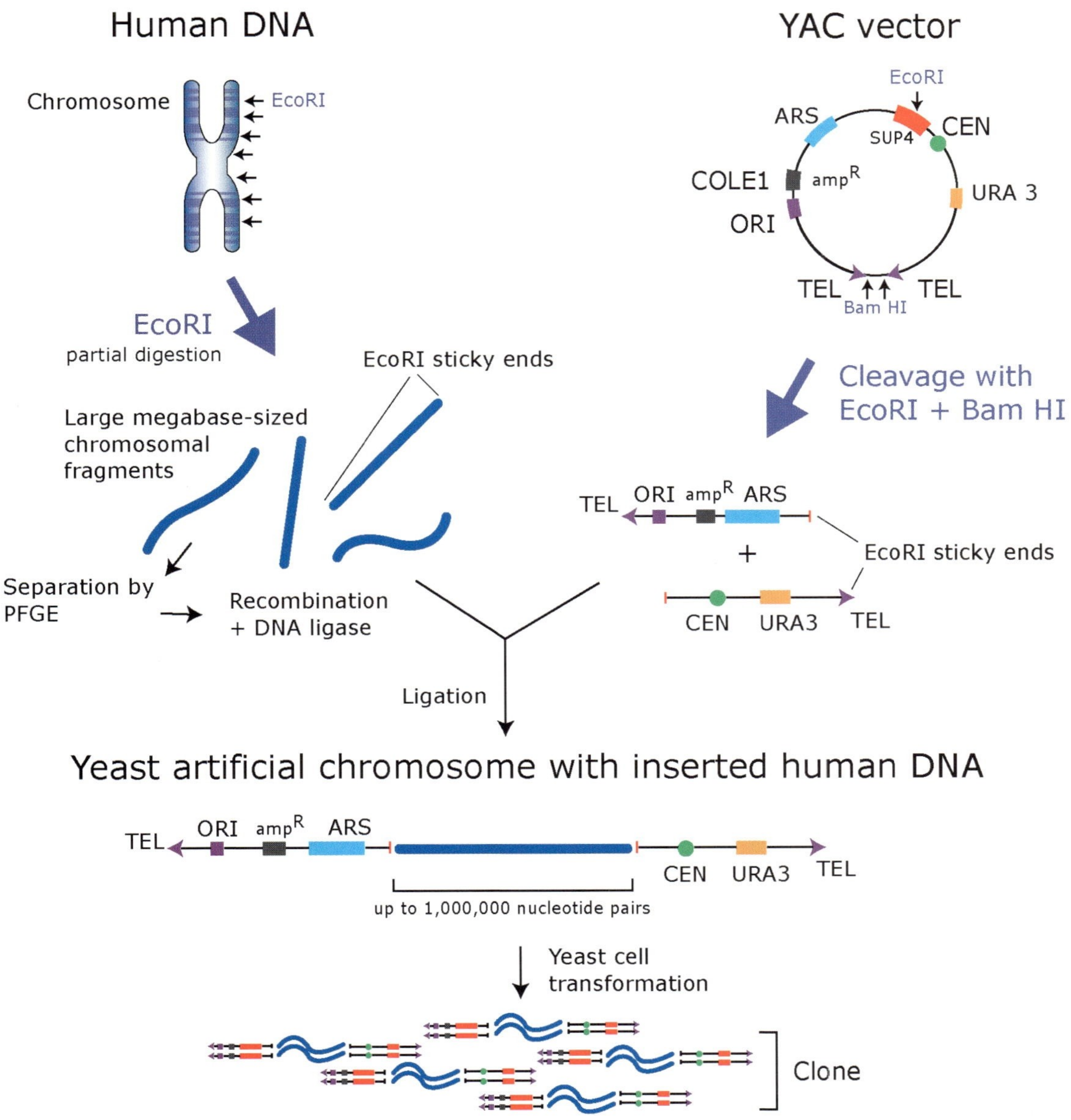

clone very large exogenous DNA fragments has made YACs a widely used tool in physical mapping. The length of target DNA insert can be as large as two megabases. The YAC vectors have, therefore, become the major workhorses of the Human Genome Project.

Suggested Reading

Giraldo, P. and L. Montoliu. 2001. Size matters: use of YACs, BACs and PACs in transgenic animals. Transgenic Res. *10*:83-103.

REFERENCE TEXTS

T. Strachan and A.P. Read. 1999. Human Molecular Genetics, 2nd edition. Wiley-Liss, New York (ISBN 0471330612).

R.L. Miesfeld. 1999. Applied Molecular Genetics, 1st edition. Wiley-Liss, New York (ISBN 0471156760).

D.H. Gelfand, J.J. Sninsky, T.J. White, and M. Innis (Eds.). 1990. PCR Protocols: A Guide to Methods and Applications. Academic Press, San Diego (ISBN 0123721814).

B. Alberts, A. Johnson, J. Lewis, M. Raff, K. Roberts, and P. Walter. 2002. Molecular Biology of the Cell. Garland Publishing, 4th edition (ISBN 0815332181).

H. Lodish, A. Berk, S.L. Zipursky, P. Matsudaira, D. Baltimore, and J. Darnell. 1999. Molecular Cell Biology, 4th edition. W.H. Freeman and Company, New York (ISBN 0716731363).

M.R. Walker and R. Rapley. 1997. Route Maps in Gene Technology, 1st edition. Blackwell Science, Williston, VT. ISBN: 063203792X.

J.M. Walker and R. Rapley. 2001. Molecular Biology and Biotechnology, 4th edition. Royal Society of Chemistry, London (ISBN 0854046062).

Index

E

F

G

Q

R